High-Purity Metals and Alloys
Fabrication, Properties, and Testing

High-Purity Metals and Alloys
Fabrication, Properties, and Testing

Edited by V. S. Emel'yanov
and A. I. Evstyukhin

Translated from Russian

Springer Science+Business Media, LLC 1967

ISBN 978-1-4899-4730-7   ISBN 978-1-4899-4728-4 (eBook)
DOI 10.1007/978-1-4899-4728-4

The original Russian text was published for the Ministry of Higher and
Intermediate Specialized Education of the RSFSR and the Moscow Engineering Physics Institute by Atomizdat, Moscow, in 1966.

В. С. ЕМЕЛЬЯНОВ и А. И. ЕВСТЮХИН
МЕТАЛЛУРГИЯ И МЕТАЛЛОВЕДЕНИЕ ЧИСТЫХ МЕТАЛЛОВ
METALLURGIYA I METALLOVEDENIE CHISTYKH METALLOV

Library of Congress Catalog Card Number 67-19386

PREFACE

The majority of the twenty-six articles in this collection were contributed to scientific conferences of institute teachers in 1963 and 1964 ("Metallurgy and Metallography" Section). The articles are devoted to developing methods of producing pure metals and alloys used in modern technology and to studying their properties. In some of the articles new apparatus for studying the physical properties of metals is described.

As in the previous issues, progress in the application of the iodide and chloride methods of refining refractory metals (especially vanadium) is described, together with improvements to the zone-melting method.

In order to elucidate the nature of the interaction between zirconium and various impurities and alloying elements (hydrogen, carbon, yttrium, hafnium, and niobium), the structures and physical properties of systems containing these elements were examined. The distribution of hydrogen in the presence of a temperature gradient is extremely important from the point of view of reactor use. This question is considered in one of the experimental papers published in the collection. The use of high-alloy zirconium-base material as a superconductor with high critical parameters is also of interest, and one paper is devoted to this question.

Questions of niobium refining and the structure of alloys giving corrosion-resistant films on niobium are discussed in several papers.

A second group of papers is concerned with self diffusion in pure uranium, the effect of purity on the behavior of uranium during annealing, and the compatibility of uranium carbide with tungsten. This group also includes papers on methods of studying the diffusion coefficient for α-radioactive elements as well as more general questions in the theory of diffusion.

Three papers are devoted to the mechanism underlying the interaction of molten metallic lithium with construction materials.

The third group of articles describes experimental systems developed by members of the faculty; these include apparatus for the extrusion of metals in liquid form under high pressure, a system for measuring hot hardness by static methods, equipment for floating-zone melting, an adaptation for an x-ray camera, and so forth.

The contents of this collection are of special interest to a wide circle of technical engineers, students, and graduates concerned with the study and development of new materials.

SOVIET ALLOY DESIGNATIONS

The alloy designations in this volume have been transliterated from the Russian rather than translated to avoid confusion with Western alloys that are similar but not identical. The explanation of these designations that follows is based on the one contained in the Handbook of Soviet Alloy Compositions (U.S. Department of Commerce, OTS, Washington 25, D.C., PB 171331) and the British Iron and Steel Industry Translation Service Publication No. BISI 2000. A complete description of steel and alloy compositions is given by Soviet steel standards, GOST 4543-57 and 4632-51.

Constructional carbon steels of "ordinary" quality, which are specified according to their mechanical properties, are St0, St1, St2, St3, St4, St5, St6, and St7. The higher the figure, the greater the carbon content. Depending on the method of manufacture, the prefix B or M may be added to describe a Bessemer or open-hearth steel, respectively.

Constructional "quality" carbon steels have numbers between 05 and 70, which indicate the average carbon content in hundredths of one percent. These numbers may carry suffixes kp, sp, and ps, denoting rimmed, killed, and semikilled grades, respectively.

"Quality" carbon tool steels have designations U7, U8, etc., up to U12, the number indicating the average carbon content in tenths of one percent.

In alloy steel designations, the following symbols represent the alloying elements:

A	Nitrogen	Kh	Chromium	S	Silicon
B	Columbium	M	Molybdenum	T	Titanium
D	Copper	N	Nickel	Ts	Zirconium
F	Vanadium	P	Phosphorus	V	Tungsten
G	Manganese	R	Boron	Ya	Aluminum
K	Cobalt				

If the percentage of the element is not greater than about 1%, the letter for the element is not followed by a figure. If the amount of the element is greater than 1%, a figure representing the content is placed to the right of the letter, e.g., 4% Ni is represented by N4.

The average carbon content is shown to the left of the letters as hundredths of one percent. In the case of a very low carbon content (less than about 0.08%), the numeral 0 is placed before the letters. Occasionally, the carbon figure is omitted altogether.

The letter A appended to these designations denotes "high quality" (narrower composition limits and lower sulfur and phosphorus contents). This must not be confused with the symbol A representing nitrogen. Frequently, the letter L is used as a suffix to denote a cast steel.

Several groups of steel designations carry a prefixed letter which indicates a particular purpose or characteristic of the steel, e.g., A free-machining, E magnetic, Zh straight chromium stainless, R high-speed, Sh ball-bearing, E electrical, and Ya chrome-nickel stainless. The numbers following these letters are not normally indicative of the actual composition.

Many steels and alloys have designations consisting of the letters ÉI or EP followed by a number of up to three digits. Wherever possible, the actual compositions of these materials will be given.

CONTENTS

IODIDE-TYPE VANADIUM

V. S. Emel'yanov, A. I. Evstyukhin,
V. I. Statsenko, and A. I. Dashkovskii

Pure vanadium is ductile and has a high melting point ($T_m = 1900 \pm 25°C$). The specific gravity (6.1 g/cm^3) is 35% higher than that of titanium. As regards heat resistance, vanadium-titanium alloys constitute the best material for working temperatures of 350 to 600°C. It is well known that vanadium sheet may be used for aviation components; such sheet suffers negligible oxidation in air up to temperatures of 500°C, even in the absence of protective coatings, and is quite compatible with organic fuel. Vanadium is extremely resistant to corrosion in highly reactive chemical media, which makes it almost irreplaceable as a material for chemical apparatus. This metal is also promising for use in nuclear technology, being compatible with uranium and having a capture cross section of 4.7 to 5.1 b. Reactors in the United States use thin-walled inner tubes of vanadium operating under pressure at 300 to 400°C. Such tubes can be successfully electron-beam welded. For fast-neutron reactors, coating the fuel elements with vanadium is of special interest, since the metal is completely compatible with uranium and has high thermal conductivity as well as corrosion resistance. Vanadium may clearly find uses in other branches of technology as well.

The use of metallic vanadium has nevertheless been limited up to the present time by the difficulty of producing it in pure form. For the same reason the physicochemical properties of vanadium and its heat resistance have not been adequately studied. In this paper we describe experiments in preparing pure vanadium by the iodide method from Soviet produced aluminothermal and carbothermal metal. Bars of iodide-type vanadium were remelted in arc and electron beam furnaces and then subjected to working. Properties of the pure metal (internal friction and shear modulus at various temperatures) were studied in the worked samples, and interstitial impurities such as oxygen, nitrogen, and hydrogen in iodide vanadium were determined.

Preparation of Iodide Vanadium and Its Properties

It was shown in earlier issues of this collection [1] that the greatest purification of refractory metals from nonmetallic impurities such as carbon, oxygen, nitrogen, and hydrogen can be achieved by thermal decomposition of their volatile compounds.

In the present investigation, iodide vanadium was obtained in a quartz apparatus provided with induction heating, analogous to that described in [1], and also in apparatus made of refractory glass with a filament heated by the passage of current through electrodes sealed into the apparatus.

Table of the Average Microhardness of Commercial
and Iodide-Refined Vanadium

Form of metal	Micro-hardness, kg/mm^2
Original:	
aluminothermal	645
carbothermal	183
Iodide:	
from aluminothermal metal. . .	88–90
from carbothermal metal	72
remelted in arc furnace	76
worked	156
remelted in electron beam . . .	70
from [2]	64

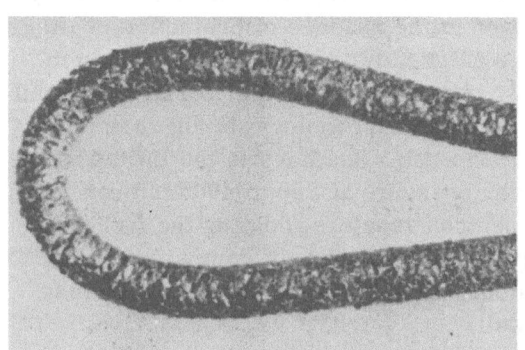

Fig. 1. External form of an iodide-vanadium rod (×2).

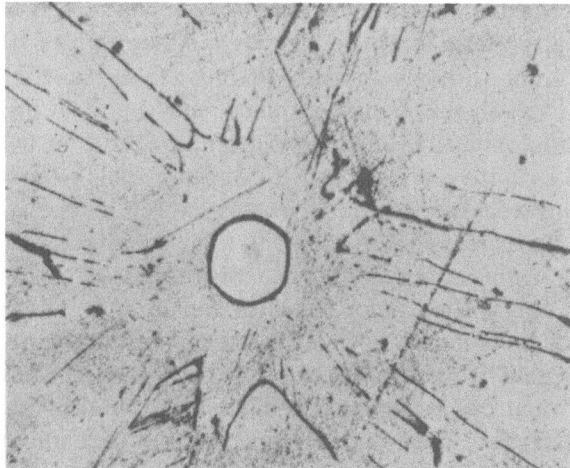

Fig. 2. Microstructure of an iodide-vanadium rod (×110).

The original metal for refining was commercial vanadium obtained by the aluminothermal and carbothermal methods. This metal was loaded into the apparatus in the form of small pieces, degassed by heating to a residual pressure of 10^{-5} mm Hg, and disconnected from the vanadium system. Then iodine vapor was introduced into the apparatus, and on further heating this reacted with the metal, forming iodides.

According to [2] vanadium forms two compounds (stable in the condensed phase) with iodine: VI_2 and VI_3. At temperatures between 280 and 400°C, VI_3 decomposes completely into VI_2 and free iodine. Morette [3] found that VI_2 sublimed at 750°C and underwent partial thermal decomposition at 800°C.

As a result of our own investigations, we found that the vanadium began to be deposited from the gas phase onto a heated (1100°C) tungsten wire when the temperature of the reaction vessel (and hence the raw material) was approximately 475°C. Certainly the rate of deposition was very low at this temperature; on average it equalled 0.025 g/h, but even this was much higher than the value given in [4].

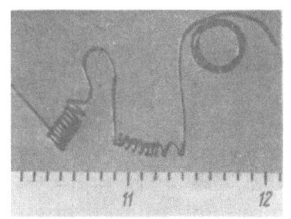

Fig. 3. Iodide-vanadium
wire.

The rate of vanadium deposition increased with the temperature of the reaction vessel up to 820°C; above this the process could no longer be followed, owing to the sublimation and deposition of vanadium on the walls of the vessel and the insulators supporting the tungsten filament, which caused the process to cease.

The temperature of the filament was kept between 1100 and 1200°C in all the experiments.

Our experiments indicated that the optimum temperature of the reaction vessel for the iodide refining of vanadium was 800°C. The rate of vanadium deposition at this temperature reached 1.8 g/h. In this way we obtained very dense and ductile bars of iodide-type vanadium, the external form and microstructure of which are shown in Figs. 1 and 2. The average microhardnesses of the original and resultant vanadium, which qualitatively characterize the degree of purification from nonmetallic inclusions, are shown in the table.

We see from the table that the microhardness of aluminothermal vanadium, originally 645 kg/mm^2, fell to 90 kg/mm^2 (a factor of 7.2) after iodide refining; the microhardness of the carbothermal metal fell from 183 to 72 kg/mm^2 (factor of 2.5).

After remelting the iodide vanadium in an arc furnace, the microhardness rose slightly (76 kg/mm^2), and after electron-beam melting fell to 70 kg/mm^2.

Working the iodide-type vanadium caused it to harden, the microhardness of the sample rising to 156 kg/mm^2 (almost double). The ductility of the iodide vanadium nevertheless remain so high that wire 0.2 mm in diameter could easily be prepared from it; an example is shown in Fig. 3.

In view of the low accuracy of direct methods of determining small traces of impurities, we used the internal-friction method in order to determine these in iodide vanadium; this method enables the shear modulus and other elastic properties of the metal to be found.

Internal Friction in Iodide Vanadium

The analysis of impurities in iodide vanadium is an extremely complex problem. It is well known, however, that the presence of even minute traces of oxygen and nitrogen in vanadium leads to the appearance of internal-friction peaks due to the diffusion of these atoms in the stress field [5, 6]; the height of the peaks increases as the oxygen and nitrogen content grows.

We measured the internal friction of our iodide-vanadium bars on the apparatus described earlier [7], using torsional oscillations. The internal friction was measured on heating and cooling at an average rate of approximately 100 deg/h (up to 400°C) and 250 deg/h at higher temperatures. Isothermal rests were made during the measurements and the amplitude dependence of the internal friction was determined over the deformation-amplitude range 1.5×10^{-4} to 1.65×10^{-5}. Over the whole temperature range (20 to 400°C) the amplitude graphs were straight lines almost parallel to the x axis, rising slightly with increasing amplitude and practically parallel to each other. The value of internal friction corresponding to the minimum amplitude (1.65×10^{-5}) for each temperature was plotted on the graph relating internal friction to temperature. The temperature dependence on the shear modulus was plotted as a function of the ratio of the squares of the oscillation frequencies at the measuring temperatures and room temperature, respectively.

The temperature relationships for the internal friction and shear modulus of a sample forming part of an iodide-vanadium bar 1.8 mm in diameter and 180 mm long are shown in

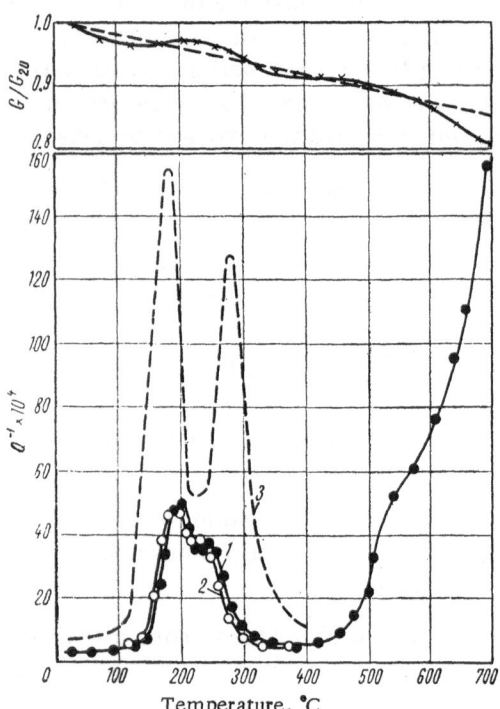

Fig. 4. Internal friction and shear modulus of iodide vanadium as functions of temperature: 1) Heating; 2) cooling; 3) calcium-thermal vanadium, according to [5].

Fig. 4. Curves 1 and 2 show two smallish, partly overlapping peaks, the first at 195°C and the second (lower) one at 255°C. The two peaks are satisfactorily reproducible on both heating and cooling the sample.

The temperatures at which these peaks are observed, as well as their shape and mutual disposition, entirely correspond to the internal-friction peaks arising in vanadium as a result of the stress-induced diffusion of oxygen and nitrogen atoms [5, 6].

The height of the peaks in iodide vanadium, however, as well as the height of the internal-friction background, were much lower than those, for example, in vanadium obtained by the calcium-thermal method [5], the data for which are shown by curve 3 in Fig. 4.

The first peak at 195°C is evidently due to the presence of interstitial oxygen atoms in the crystal lattice of the vanadium. If we consider the displacement of the temperature peak for the oscillation frequency at which the measurements were made and use the activation energy of this peak (28.2 to 28.6 kcal/mole) obtained in [5, 6], we find that at an oscillation frequency of 1.75 cps the peak due to dissolved oxygen atoms should occur at a temperature near 195°C.

The second peak is probably due to the presence of nitrogen in the vanadium lattice; despite the high oscillation frequency used in the present experiment, it occurs at a lower temperature (255°C) than that indicated in [5, 6]. This behavior, however, may be due to the rather small amount of nitrogen in the vanadium, since it is well known that the temperature of the peak depends to a certain extent on the quantity of dissolved impurity; increasing dissolved impurity content broadens the peak and moves it in the high-temperature direction. In the present case the small height of the peak also indicates a small amount of nitrogen present.

The presence of hydrogen in vanadium produces relaxation peaks, the height of which depends on the hydrogen concentration. According to [8], the internal-friction background increases even for 1.2 at.% hydrogen; for large quantities of hydrogen a peak develops at temperatures under 120°C, especially when the samples are being heated. The absence of any rise in internal friction near room temperature in iodide vanadium indicates that, even if hydrogen is not entirely absent, its concentration is certainly lower than 1.2 at.%.

The internal friction curve of iodide vanadium shows a slight peak at approximately 550°C (Fig. 4, curve 2); this was not found in [5, 6]. This is probably associated with relaxation at grain boundaries, which becomes possible in the purer material. The extent of this relaxation effect, however, is not great, since the iodide rods studied had large grain sizes.

The shear modulus of iodide vanadium falls nonlinearly with increasing temperature, and in the temperature range corresponding to the internal-friction peaks there is even a slight rise. The average temperature coefficient of the shear modulus of iodide-type vanadium is not great; at 600°C the modulus is 12% smaller than at room temperature. At temperatures above 600°C the fall in the modulus takes place more sharply.

Conclusions

1. We have developed a method for the iodide refining of vanadium and determined the optimum conditions for the process.

2. We have obtained compact iodide-type vanadium with a very high density and a ductility sufficient to enable wire 0.2 mm in diameter to be prepared from it without intermediate annealing.

3. We have measured the internal friction and shear modulus of iodide vanadium as a function of temperature. The curves so obtained enable a semiquantitative estimate of the concentration of interstitial impurities (oxygen, nitrogen, hydrogen) to be made by a comparison technique.

Literature Cited

1. Metallurgy and Metallography of Pure Metals. Collection of articles edited by V. S. Emel'yanov and A. I. Evstyukhin. No. 1, Moscow, Izd. MIFI, 1959; No. 2, Moscow, Atomizdat, 1960; No. 3, 1961; No. 4, 1963. Moscow, Gosatomizdat.
2. Tolmacheva, T. A., et al. Zh. Neorgan. Khim., 8:553 (1963).
3. Morette, A. Compt. Rend., 207:1218 (1938).
4. Carlson, O. N., and Owen, C. V. J. Electrochem. Soc., 108(1):88 (1961).
5. Powers, R. W. Acta Met., 2:604 (1954).
6. Stanley, J. T., and Wert, C. A. Acta Met., 3:107 (1955).
7. Dashkovskii, A. I., et al. In collection: Metallurgy and Metallography of Pure Metals, No. 2, Moscow, Atomizdat, 1960, p. 207.
8. Butera, R. A., and Rofstad, P. J. Appl. Phys., 34:2172 (1963).

EXPERIMENTAL STUDY OF THE EFFECT OF TEMPERATURE GRADIENT ON THE DISTRIBUTION OF HYDROGEN IN ZIRCONIUM

V. S. Emel'yanov, N. V. Borkov, A. I. Evstyukhin, and A. T. Kazakevich

One of the problems associated with the use of zirconium or its alloys as a structural material in water-cooled nuclear reactors is the active absorption of hydrogen by zirconium during the operation of the reactor. The hydrides thus formed cause deterioration in the mechanical and anticorrosive properties of components made from zirconium and its alloys.

Some components of the reactor core, such as the jackets (cans) of the fuel elements, operate with large temperature drops between the inner and outer surfaces. As we know [1-3], a temperature gradient causes the hydrogen in zirconium to migrate from a heated surface to a cold one. This results in a concentration of hydrogen and the formation of hydrides at the cooled surface of the can.

We therefore made an experimental study of the redistribution of hydrogen in Soviet-produced cylindrical zirconium samples with temperature gradients along the length and radius of the cylinder. The experiments were based on metallographic and chemical methods of analysis.

A special apparatus was designed to establish a temperature gradient in the cylindrical zirconium samples. The basic drawing of the apparatus is shown in Fig. 1 and a general view in Fig. 2.

Sample 1 was designed for establishing a temperature gradient over the radius; the sample constituted a hollow cylinder 1545 g in weight. The length of the cylinder was 132 mm, the outside diameter 45.2 mm and the inside diameter 9.8 mm. A copper tube was embedded in the central aperture of the sample (Fig. 3).

Hydrogen (0.05 wt.%) was first introduced into the sample by the technique described in [4] and the sample was subjected to homogenization at a temperature of 660°C for a period of 34 h. Metallographic analysis indicated that, after this treatment, the hydrides were uniformly distributed over the whole section (Fig. 4).

During the experiment, the sample was held in the furnace in a vertical position. To ensure thermal insulation, the ends of the furnace were covered with asbestos plates; the ends of the copper cooling tube were brought out through these plates. Thus, by heating the outside of the sample in the furnace and cooling the inside with water, we established a temperature gradient over the radius of the sample. To measure the temperature at different points of the sample, five thermocouple holes were drilled at each end. The thermocouples were uniformly

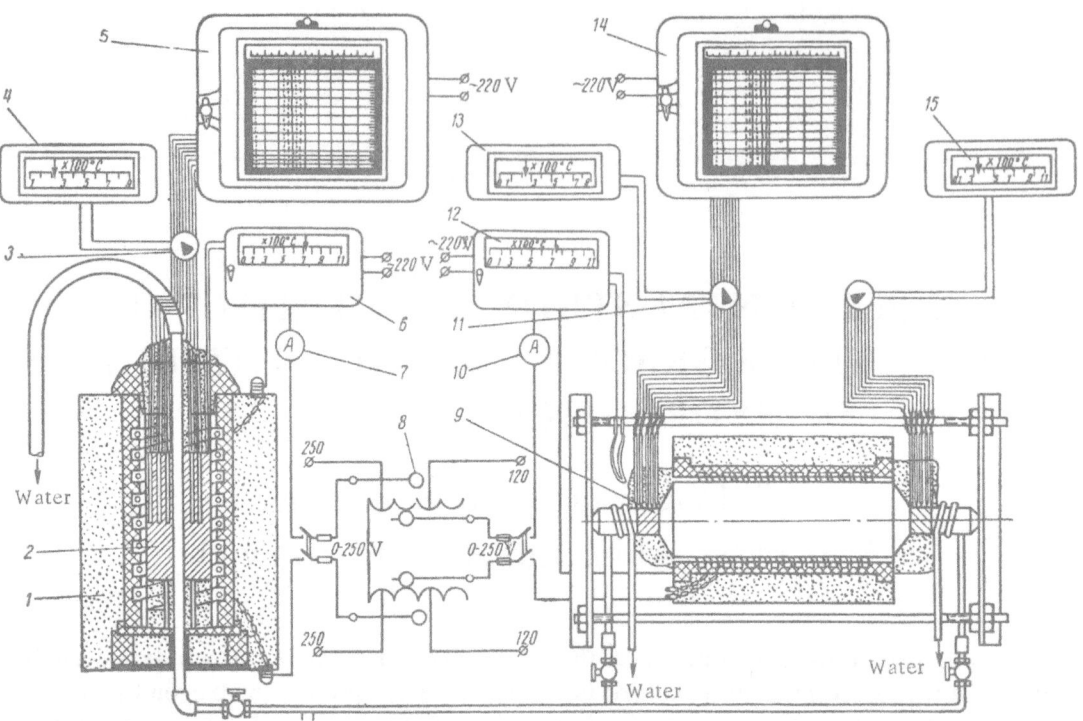

Fig. 1. Diagram of apparatus for obtaining a temperature gradient in cylindrical zirconium samples: 1) TG-1 furnace; 2) sample 1; 3, 11) thermocouple switch; 4, 13, 15) galvanometers; 5, 14) recording instruments (ÉPP-09-M1); 6, 12) furnace temperature controllers; 7, 10) ammeters; 8) transformer; 9) sample 2.

spaced at intervals of 3.2, 6.2. 9.2, 12.2, and 15.2 mm from the inner surface; this made it possible to determine the temperature gradient over the radius of the sample experimentally and to keep it constant during the course of the experiment. The temperature-inside the TG-1 furnace was 700°C for a current of 14 A and a power of 2.7 KVA.

Data relating to the flow of cooling water and the thermal conditions of the sample are indicated in Table 1.

The experiment ran continuously for 740 h. The desired (steady) thermal conditions were attained 1 h after the apparatus had been switched on and remained unchanged for the duration of the experiment.

After the furnace heater had been switched off, sample 1 was removed from the furnace and cooled to room temperature. Then it was cut into segments, from which microsections were made. The microstructure of a typical microsection cut in a radial direction is shown in Fig. 5.

Metallographic examination of a zirconium sample containing 0.05 wt.% hydrogen indicated that, when a temperature gradient acts for a long period, the distribution of the hydride phase over the radius of the sample is not uniform. In the region to the left of the microsection (the part touching the water–cooled inner surface of sample 1, Fig. 5), the hydride phase increased; in the region to the right (the part touching the heated outer surface), the hydride phase became less abundant than that in the original microsection (Fig. 4).

Fig. 2. Apparatus for the establishment of a temperature gradient in cylindrical zirconium samples.

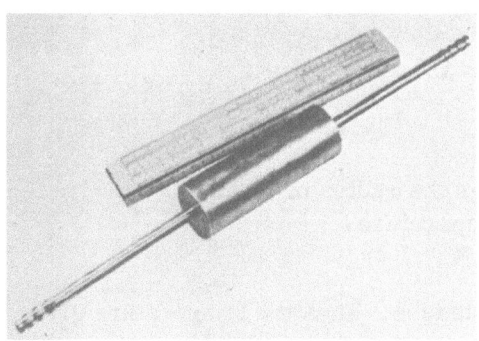

Fig. 3. Cylindrical zirconium sample 1 with central copper tube .

This variation in hydrogen concentration over the radius of the sample was confirmed by the chemical analysis of lathe chips taken from the center section of the sample. Forty-three such samples were examined.

Figure 6 shows the distribution of hydrogen along the radius of the sample as a function of temperature. The straight line 1 illustrates the uniform distribution of hydrogen over the radius existing before the experiment (hydrogen content 0.05 wt.%). At room temperature, most of the hydrogen exists as a chemical compound (hydride). Curve 2 shows the temperature variation over the radius after steady operating conditions have been established. The surface temperatures (260 and 430°C) were determined by extrapolating the curve plotted from the thermocouple readings.

Curve 3 shows the hydrogen distribution over the radius of the sample after the completion of the experiment. This curve was plotted from the results of the chemical analysis and metallographic examination; it shows that the hydrogen content is smallest at the heated (outer) surface of the sample (0.0094 wt.%). The cooled (inner) surface of the sample had the greatest hydrogen concentration (0.14 wt.%).

Knowing the temperature as a function of radius (curve 2, Fig. 6), we can determine how much hydrogen is in solid solution at a given point in the sample and at a given temperature.

We used the formula $C = 8.5 \exp(-7600/Rt)$, taken from Sawatsky's paper [5], to make an approximate determination* of the solubility of hydrogen in α-zirconium as a function of the radius of the sample at the operating temperatures (see curve 4, Fig. 6).

The initial hydrogen concentration (0.05 wt.%) and its uniform distribution over the sample cross section indicate the presence of two phases over the whole radius of the cylinder — a solid solution of hydrogen in α-zirconium and hydride phase. At room temperature, the two phases were in thermodynamic equilibrium. After the establishment of the operating temperature (see curve 2, Fig. 6), some of the hydrides dissolved in accordance with curve 4; however, the two-phase structure remained over the entire cross section of the sample for some time. The second phase (hydride) was not uniformly distributed over the radius of the cylinder; the heated

---

*Curve 4 is an approximation, since the formula given in [5] was intended for Zircalloy-2.

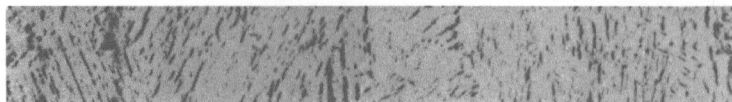

Fig. 4. Microstructure of sample 1. The microsection
was cut in a radial direction before applying a temperature
gradient. The hydride impurities can be seen scattered over
the microsection.

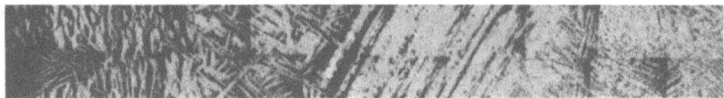

Fig. 5. Microstructure of sample 1, after imposition
of the temperature gradient. The microsection was cut in
a radial direction. On the left is the cooled (inner) surface of the cylinder; on the right is the heated (outer) surface. The hydride phase in the heated region is less abundant than that of the original microsection.

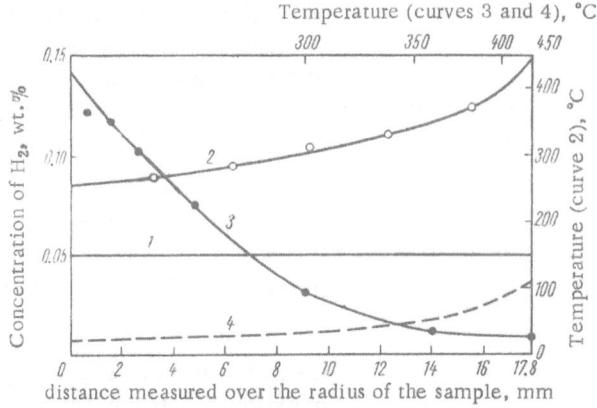

Fig. 6. Hydrogen distribution over the radius of
sample 1 as a function of temperature.

surface contained less of this phase than did the cooled surface. Thus the temperature gradient caused the hydrogen to diffuse into a colder region. To restore equilibrium between the
hydride phase and the solid solution, hydrogen began passing into solid solution. This process continued until the hydrides were completely dissolved in the hot part of the sample. At
a temperature of 430°C, up to 0.035 wt.% of hydrogen dissolved in α-zirconium. After some
time, only one phase remained in the region near the outer surface of the sample, namely, a
solid solution of hydrogen in α-zirconium.

The interface between the single-phase and two-phase regions gradually shifted from
the hot surface to the cold surface. This shift continued until equilibrium was established between the flow of hydrogen from the hot surface to the cold (resulting from the temperature
gradient) and the flow of hydrogen from the cold surface to the hot (resulting from the concentration gradient). This equilibrium state is represented by a particular hydrogen-concen-

TABLE 1. Characteristics of Thermal Conditions for Sample 1

Parameter	Quantitative characteristics
Coolant flow rate	30 ml/sec
Pressure of water supply system	2.5 atm
Water temperature on entering sample	4°C
Water temperature on leaving sample	16°C
Temperature at the outer surface of the sample	430°C
Temperature at the inner surface of the sample (extrapolation)	260°C
Temperature gradient at the outer (heated) surface of the sample...............	250 deg/cm
Temperature gradient at the inner (cooled) surface of the sample...............	35 deg/cm
Thermal flux through the sample	475 cal/sec
Thermal flux through the heated surface ...	2.5 cal/cm^2·sec
Thermal flux through the cooled surface ...	11.5 cal/cm^2·sec

TABLE 2. Thermal Conditions for Samples 2 and 3

Parameter	Quantitative characteristics
Water–coolant flow rate	20 ml/sec
Water temperature at inlet	4°C
Water temperature at outlet...........	6°C
Temperature of the heated end of the sample	383°C
Temperature of the cooled end of the sample	153°C
Heat flux through the sample	6.5 cal/cm^2·sec

tration radius relationship (see curve 3, Fig. 6). The interface between the single-phase and two-phase regions lay at the intersection between curves 3 and 4 after 740 h, that is, at a distance of 13 mm from the inner surface of the sample; this corresponds to a hydrogen content of 0.015 wt.% and a temperature of approximately 350°C.

After the establishment of the operating temperatures, the region to the right of the interface became single-phased (solid solution of hydrogen in α-zirconium), and the region to the left two-phased (solid solution and hydrides). However, after the experiment had ended and the sample had cooled to room temperature, the limiting solubility of the hydrogen in the zirconium decreased; this caused the hydride phase to precipitate over the entire cross section of the sample. The single-phase region cannot therefore be seen in Fig. 5.

Cylindrical samples 2 and 3, supporting longitudinal temperature gradient, were tested at the same time as sample 1 and on the same apparatus. Each of these samples was first saturated with hydrogen (0.01 wt.%). The samples were then subjected to homogenizing at 660°C for 4 h; this produced a uniform distribution of the hydrides over the whole cross section.

To obtain a temperature gradient along the length of the cylinder, one end was heated and the other cooled. This was accomplished in the following way. A massive stainless-steel

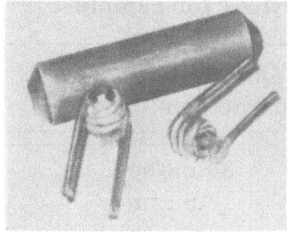

Fig. 7. Steel core and copper cylinders carrying cooling coils.

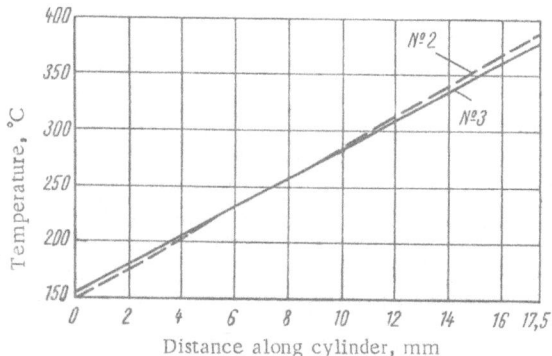

Fig. 8. Temperature of samples 2 and 3 as a function of distance along the cylinder under steady operating conditions.

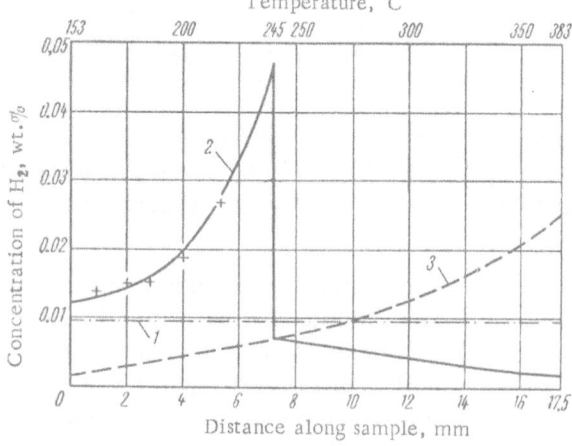

Fig. 9. Hydrogen concentration along the cylinder as a function of temperature (samples 2 and 3): 1) Initial hydrogen concentration at different points of the sample; 2) hydrogen concentration at different points of the sample after the completion of the experiment; 3) limiting solubility of hydrogen in α-zirconium as a function of temperature.

core was inserted into a cylindrical furnace set up in a horizontal position. Copper tube coils were wound around small copper cylinders (Fig. 7) and soldered. The ends of the core, the small copper cylinders, and the zirconium samples were carefully polished. The experimental conditions and results were much the same for both samples. The results given below relate only to sample 2.

The weight of the sample was 34.3 g, the diameter 20 mm, and the length 17.5 mm. In the operating position, one end of each sample was pressed tight against the steel core inside the furnace. The other ends of the samples were in good thermal contact with the copper cooling cylinders. Table 2 indicates the thermal conditions for a furnace temperature of 700°C with water cooling of the copper cylinders.

Each sample contained five thermocouples connected to a recording instrument. The thermocouples were placed along the length of the cylinder at the following distances from the heated end: 3.5, 5.0, 7.5, 12.5, and 15.0 mm. The temperature of the samples was kept constant throughout the 740-h experiment. The temperatures along the length of sample are given in Fig. 8.

The data presented in Table 2 and Fig. 8 indicate that the temperature drop between the ends of each sample was 230°C. The temperature gradient had a constant value of 130 deg/cm.

After the furnace had been switched off, the samples were removed from the core and cooled in air; metallographic examination and chemical analysis followed. Twenty-four specimens were cut from each sample for chemical analysis.

As a result of the experiment, the hydrogen was redistributed longitudinally: in the more heated regions the concentration fell below its original valve, and in the less heated regions it increased. The variation in hydrogen concentration along the sample is shown in Fig. 9. The intersection between curves 2 and 3 indicates the interface between the single-phase and two-phase regions after the establishment of steady temperature conditions. At this point (7.2 mm from the cooled end) the hydrogen concentration was greatest. The corresponding operating temperature was 245°C.

Conclusions

1. An apparatus for obtaining longitudinal and radial temperature gradients in cylindrical samples' height has been described.

2. Under the influence of temperature gradients, the hydrogen in zirconium migrates from hot regions to cold regions.

3. The curves representing the change in hydrogen concentration due to temperature gradients as a function of position indicate that, under the conditions specified, maximum hydrogen concentration occurs where the temperature lies between 245 and 260°C.

Literature Cited

1. Spalaris, C. N., et al. Nucl. Sci. Eng., 8:83 (1960).
2. Sawatsky, A. J. Nucl. Mater., 2:321 (1960).
3. Markowitz, J. M. Progr. Nucl. Energy, Ser. V, 3:95 (1961).
4. Emel'yanov, V. S., and Borkov, N. V. In collection: Metallurgy and Metallography of Pure Metals, No. 4, Moscow, Gosatomizdat, 1963, p. 18.
5. Sawatsky, A. J. Nucl. Mater., 2:62 (1960).

DIFFUSION OF YTTRIUM IN ZIRCONIUM

G. B. Fedorov, F. I. Zhomov, and E. A. Smirnov

The self-diffusion of zirconium in the α and β phases was considered in [1–3]. A study of the diffusion mobility of yttrium in zirconium is also of great interest. Yttrium stands next to zirconium in the Periodic Table, Zr crystallizing in the hexagonal lattice, but the atomic radii of the two elements are quite different. The difference between the atomic radii of yttrium and β-zirconium is approximately 16%; this has a considerable effect on the range of solubility of yttrium in zirconium. The phase diagram of the Y−Zr system shown in Fig. 1 [4] indicates that the maximum solubility of yttrium is approximately 2 at.% at 1400°C.

In the present investigation we studied samples of iodide-type zirconium remelted in an arc furnace. The resultant ingots were forged in air, giving bars of rectangular cross section, ground, and cut. The samples were first subjected to homogenizing in the β phase.

The diffusion was studied by means of the radioactive isotope Y^{90} obtained by irradiating pieces of metallic yttrium. The isotope was deposited on the samples by vacuum evaporation. The thickness of the layer deposited was several tenths of a micron. The samples were joined together in pairs with the radioactive surfaces facing one another and wrapped in molybdenum foil. Diffusion annealing of the β-zirconium was carried out in vacuum furnaces of the TVV-4 type at 1100, 1200, 1260, and 1335°C; for comparison, diffusion in the α phase was studied at 800°C. The annealing periods were, respectively, 21, 84, 27, 26, and 60 h.

The diffusion characteristics were studied by the layer-removal method, the integral activity of the remaining part of the sample being measured [5]. It was borne in mind that the absorption coefficient μ of the β radiation of Y^{90} in zirconium equalled 30 cm^{-1}. The diffusion coefficients were determined graphically by constructing the relationship

$$\ln\left(\mu I_n + \frac{\partial I_n}{\partial x_n}\right) = I\left(x_n^2\right).$$

On the basis of the quantity of isotope deposited and the area bounded by the penetration curve, it was shown by a preliminary calculation that the amount of diffusing yttrium in the diffusion zone lay within the solubility limit for the temperature in question.

The calculated diffusion coefficients of yttrium in zirconium are shown in the table.

The temperature dependence of the diffusion coefficients of yttrium in zirconium shown in Fig. 2 (line 1) may be put in the form of the equation

$$D = 8 \cdot 10^{-5} \exp\left(-23,800/RT\right) \text{ cm}^2/\text{sec}.$$

Table of the Diffusion Coefficients
of Yttrium in Zirconium

D, cm²/sec	t, °C
$1.4 \cdot 10^{-8}$	1100
$2.0 \cdot 10^{-8}$	1200
$3.4 \cdot 10^{-8}$	1260
$4.3 \cdot 10^{-8}$	1335
$3.5 \cdot 10^{-13}$	800

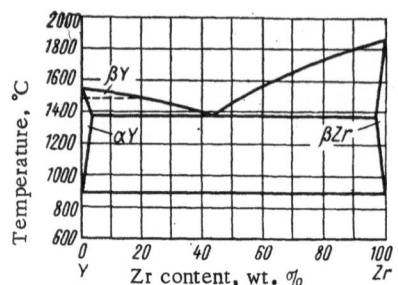

Fig. 1. Phase diagram of the yttrium-
zirconium system [4].

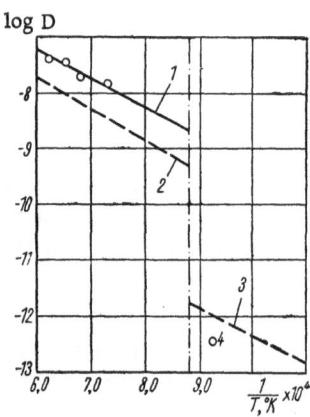

Fig. 2. Comparison between the diffusion of yttrium in zirconium and
the self-diffusion of zirconium as
functions of temperature: 1) Diffusion of yttrium in β-zirconium (our
own data); 2) self-diffusion in β-zir-
conium [2]; 3) self-diffusion in α-
zirconium [1]; 4) diffusion of yttrium
in α-zirconium (our own data).

We see that the coefficients and parameters
of the diffusion of yttrium in β-zirconium are
very close to the corresponding values for self-
diffusion in β-zirconium [2]. The diffusion coefficient of yttrium in zirconium at 800°C, which
was found for comparison, is also close to the
corresponding coefficient of self-diffusion for
zirconium in the α phase.

Conclusions

We have given the results of experiments
on the diffusion of yttrium in α- and β-zirconium.
The diffusion coefficients and parameters of yttrium are very close to the corresponding self-
diffusion characteristics of zirconium.

Literature Cited

1.   Fedorov, G. B., and Zhomov, F. I. In collection: Metallurgy and Metallography of
Pure Metals, No. 1, Moscow, Izd. MIFI,
1959, p. 162.

2.   Fedorov, G. B., and Gulyakin, V. D. In collection: Metallurgy and Metallography of
Pure Metals, No. 1, Moscow, Izd. MIFI, 1959, p. 170.

3.   Fedorov, G. B. In collection: Metallurgy and Metallography of Pure Metals, No. 4,
Moscow, Gosatomizdat, 1963, p. 34.

4.   Lurdin, C. F., and Blackledge, J. P. Presented at Joint ASMAEC Symposium on the
Rare Earths and Related Metals, Chicago, Illinois, 1959.

5.   Gruzin, P. L. In collection: Problems of Metal Science and the Physics of Metals, No. 3,
Moscow, Metallurgizdat, 1952, p. 201.

STUDY OF ALLOYS BELONGING TO THE ZIRCONIUM—CARBON SYSTEM

Yu. G. Godin, A. I. Evstyukhin, V. S. Emel'yanov, A. A. Rusakov, and I. I. Suchkov

The result of many published investigations into the properties of zirconium-carbon alloys are incomplete and contradictory.

The existence of one zirconium carbide, ZrC, has been established by a large number of authors, although it has also been suggested that there is a dicarbide [1].

Zirconium monocarbide is a phase of variable composition with a homogeneous range between 27 and 50 at.% carbon [2]. The melting point lies between 3250 and 3530°C according to various authors [3-6]. Eutectics may exist both between zirconium and the carbide [7, 4] and between the carbide and carbon [7, 8].

It has also been suggested [9, 10], however, that the interaction between zirconium and carbon has a peritectic character. The introduction of carbon into zirconium apparently raises the temperature of the zirconium $\alpha \rightarrow \beta$ transformation [7, 9].

We studied zirconium-carbon alloys by various methods of physicochemical analysis. The alloys were prepared from rods of spectroscopically-pure carbon and zirconium. We used zirconium in the form of rods obtained by the iodide method and also the powdered metal. The composition of the original zirconium is given in Table 1.

The alloys were obtained by two methods. One of these consisted of mixing zirconium and carbon powders, converting to briquettes, vacuum-sintering at 2000°C, and remelting in an arc furnace with an inert atmosphere. The other method consisted of directly fusing pieces of carbon rods and the metal in an arc furnace. This method considerably reduced the time of preparation and gave material of fairly good quality. Forty-nine alloys containing from 0.1 to 68 at.% carbon were prepared by these methods. All the alloys were analyzed chemically. It was found that, although the total carbon content in the prepared monocarbide corresponded to the stoichiometric value, the amount of combined carbon was at best 10.5 wt.% or 47.0 at.%.

The temperature corresponding to the onset of melting in the alloys was determined from the appearance of liquid in a blind opening in the center of the sample when the latter was heated by passing a current through it (the opening was made by an electron-erosion machine). The temperature was measured with a calibrated optical pyrometer (OP-48 type) graduated up to 5000°C; the measured values were corrected for the loss of radiation in the observation window of the apparatus.

Alloys in the cast state and others quenched after annealing were studied metallographically. Low-temperature annealing was carried out in evacuated quartz capsules in tubular

TABLE 1. Composition of the Original
Zirconium

Composition of bars, wt. %	Composition of powder, wt. %
Zr — 99.8; Hf — 0.04; C — less than 0.03; O — 0.03; Al — 0.02; Ni — 0.002; Fe — 0.002; Mo — 0.005; Cu — 0.001	Zr — 99.5; Fe — 0.05; Ca — 0.05; Cl — 0.002

TABLE 2. Annealing Conditions

Temperature, °C	Annealing time, h	Temperature, °C	Annealing time, h
750	250	862	200
800	250	1100	48
820	250	1700	0.5
835	200	2400	0.25
850	200		

furnaces; quenching was effected by rapidly extracting the capsules and breaking them under water.

Quenching the alloys from higher temperatures was carried out as follows: The sample was pressed between two electrodes in vacuum, heated to the required temperature, and held there for a given time; the electrodes were then moved apart, and the sample fell into D-1-A oil. The annealing conditions are shown in Table 2.

Microsections of the alloys were prepared by grinding with emery paper and polishing with a cloth moistened in a fine suspension of chromium oxide in water. Sections of satisfactory quality were obtained in this way after prolonged polishing.

The structure of the alloys was revealed by etching in a mixture of hydrofluoric and nitric acids. The HF content was very small for etching alloys with low carbon concentrations. The structure of the over-carbided alloys became quite clear in the course of polishing, and sections of these alloys needed no etching.

The alloys were studied by the x-ray powder method, using filtered copper radiation. The photographs were taken in a cylindrical RKU-86 camera, the film being placed asymmetrically. The powders for study in this way were prepared by crushing pieces of the alloys in an agate mortar. In carrying out the phase analysis, the line intensities in the majority of cases were estimated visually, using a five-point scale. In some cases we used the vanishing-phase method, in which the x-ray diffraction pictures were analyzed with an MF-4 microphotometer.

On measuring the hardness of the alloys with a diamond-tipped TP apparatus, there was a considerable spread in hardness values, owing to the great difference in the hardness of the structural components. The microhardness of the alloys was carried out on a PMT-3 hardness tester, using samples destined for microscope study.

In order to determine the solubility of zirconium in graphite, graphite powder, separated from the zirconium carbide in the over-carbided alloys containing primary graphite grains by the method described in [11], was subjected to x-ray diffraction and spectral analysis. According to the resultant data, the introduction of slight traces of carbon somewhat lowers the melting point of zirconium. On further raising the carbon content, the temperature of the onset of melting in the alloys remains practically constant at 1805°C. In alloys containing between 6.7 and 10.5 wt.% carbon (35.4 to 47.04 at.%), the temperature of the onset of melting rises rapidly and at the top of this range reaches 3500°C. Alloys containing more than 13.5 wt.% carbon (54.23 at.%) have a lower melting point, remaining almost constant at 2910°C as the carbon content rises further.

Microscope study of cast alloys in the Zr—C system showed that those containing 0.05 to 5.4 wt.% carbon (0.36 to 30.25 at.%) were two-phased. These comprised crystals of the monocarbide and grains of an α' phase constituting a supersaturated solid solution of carbon in α-zirconium. X-ray diffraction study showed that the formation of the latter was associated with the fact that the β phase was not fixed on rapidly cooling the alloys but that a phase transformation $\beta \rightarrow \alpha'$ took place. In the alloy containing 1.97 wt.% carbon (13.24 at.%), the

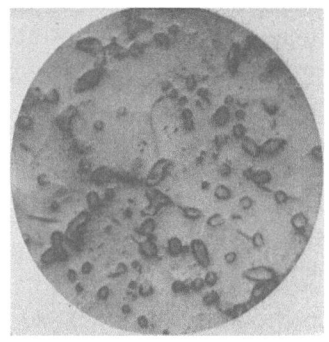

Fig. 1. Microstructure
of cast alloy containing
1.97 wt.% C (13.24 at.%).
× 180.

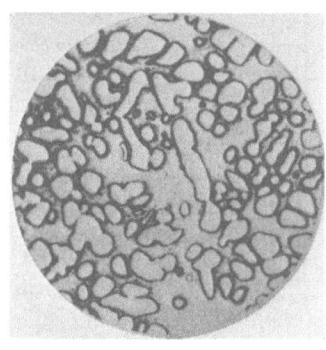

Fig. 2. Microstructure
of cast alloy containing
3.2 wt.% C (20 at.%).
× 180.

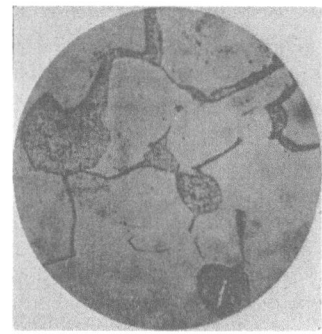

Fig. 3. Microstructure
of cast alloy containing
5.4 wt.% C (30.25 at.%).
× 180.

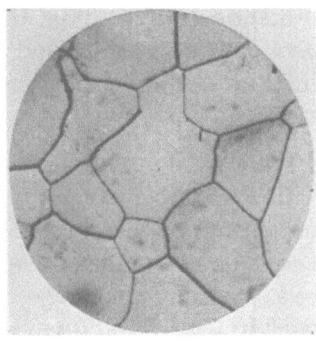

Fig. 4. Microstructure
of cast alloy containing
10.5 wt.% C (47.04 at.%).
× 180.

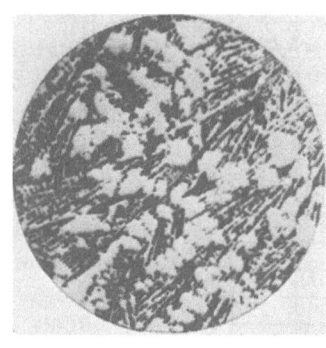

Fig. 5. Microstructure
of cast alloy containing
16.0 wt.% C (59.1 at.%).
× 180.

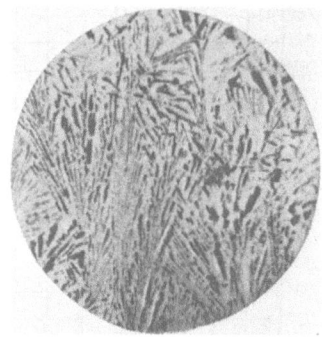

Fig. 6. Microstructure
of cast alloy containing
17.28 wt.% C (61 at.%).
× 180.

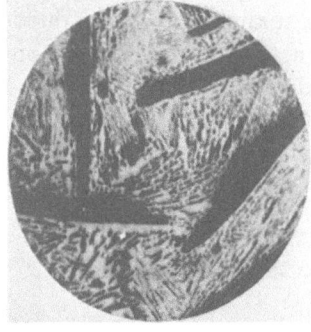

Fig. 7. Microstructure
of cast alloy containing
20.1 wt.% C (65.65 at.%).
× 180.

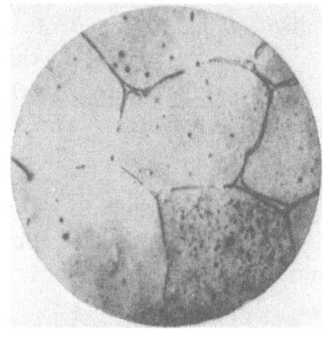

Fig. 8. Microstructure
of an alloy quenched from
1100°C, containing 6.7
wt.% C (35.3 at.%). × 180.

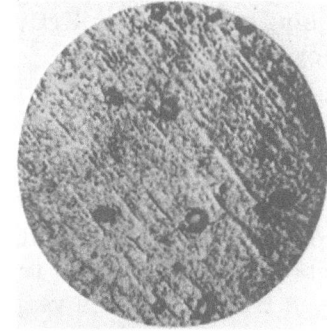

Fig. 9. Microstructure
of an alloy quenched from
1100°C, containing 0.05
wt.% C (0.38 at.%). × 180.

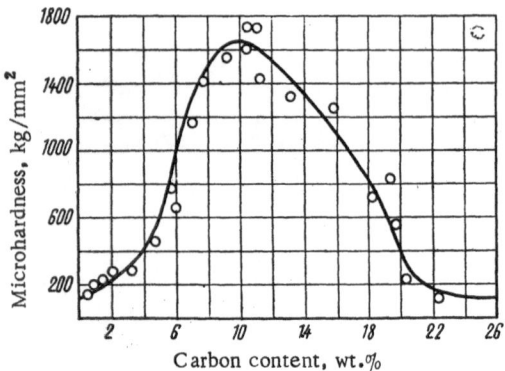

Fig. 10. Microhardness of alloys quenched from 1100°C as a function of composition.

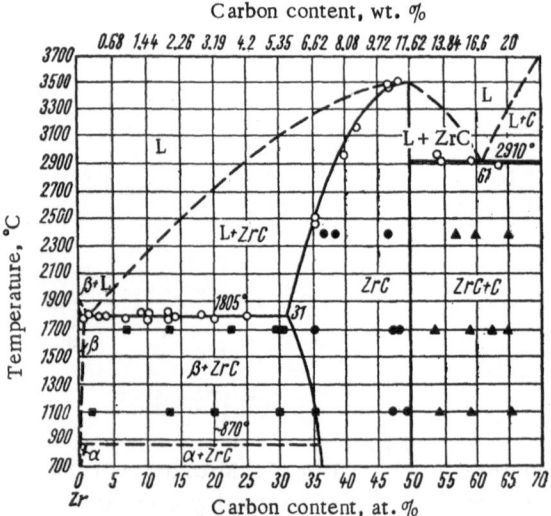

Fig. 11. Phase diagram of the Zr−C system: O) Temperature of the onset of melting in the alloys; ■) two-phase alloy β+ZrC; ●) single-phase alloy ZrC; ▲) two-phase alloy ZrC + C.

microstructure of which is shown in Fig. 1, the principal field of the section is occupied by the α' phase, on a background of which precipitates of zirconium carbide may be seen. With increasing carbon content, the quantity of carbide phase increases, and that of the α' phase correspondingly diminishes (Fig. 2); it is minimal in the alloy containing 5.4 wt.% carbon (30.25 at.%), as indicated by Fig. 3.

For carbon concentrations between 6.0 wt.% (32.66 at.%) and the composition of the monocarbide, the alloys consist entirely of the carbide phase, constituting a solid solution of zirconium in the carbide. The microstructure of an alloy belonging to this region and containing 10.5 wt.% (47.04 at.%) carbon is shown in Fig. 4. Further increasing the carbon content leads to the appearance of free graphite in the alloys, first in the eutectic (Fig. 5) existing together with the primary carbide grains, and then both in the eutectic (Fig. 6) and in the form of primary grains (Fig. 7).

Quenching alloys from 2400, 1700, and 1100°C has no serious effect on their structure, but indicates a certain fall in the solubility of zirconium in the carbide as temperature diminishes. Thus, whereas after quenching from 1700°C the alloy containing 6.7 wt.% carbon (35.5 at.%) remained single-phased after quenching from 1100°C it contained a small amount of the α phase around the grain boundaries of the carbide (Fig. 8). The solubility of carbon in the β phase of zirconium at 1100°C is below 0.1 wt.% (Fig. 9).

As the quenching temperature is reduced to 750°C, the microstructure of the alloys containing between 7.1 and 10.5 wt.% carbon constitutes a solid solution of zirconium in the carbide. The solubility of carbon in α-zirconium at 820°C is approximately 0.3 wt.%.

On quenching from 800°C or lower, alloys with small carbon contents consist of the α phase and the carbide, the amount of which rises with increasing carbon content. These facts indicate that small traces of carbon added to zirconium raise the temperature of the α → β transformations very slightly.

The results of the x-ray phase analysis confirm the microscope data, as regards the presence of a single carbide phase in the Zr−C system and the existence of a range of homogeneity.

Owing to the large spread in the lattice spacings of the alloys in the single-phase region it was impossible to determine the boundaries of the latter by means of x-ray analysis. The lattice spacing of the two-phase alloys with compositions between the carbide and zirconium has a rather higher value (4.695 A) after quenching from 2400°C than after quenching from 1700°C (4.693 A), which indicates a very slight change in the solubility of zirconium in the monocarbide as temperature falls.

Figure 10 shows the hardness/composition curve of alloys quenched from 1100°C. Analogous curves were obtained for cast alloys and those quenched from 1700°C. All the curves show a single clear maximum corresponding to the composition of the monocarbide (11.63 wt.% C). The introduction of carbon into zirconium raises its hardness until the carbide is formed; further increasing the carbon content reduces the hardness of the alloy owing to the appearance of free graphite, both in the form of the eutectic and in the form of primary crystals. No great variation in the hardness of the cast and quenched alloys was observed; this indicated an insignificant change in the solubility of zirconium in the carbide with falling temperature.

Measurements of the microhardness of the cast alloys showed that the hardness of zirconium monocarbide was 2150 kg/mm^2, varying over the homogeneous range. The microhardness of the α' phase remained equal to 180 kg/mm^2 as the carbon content increased, this being higher than the microhardness of α-zirconium (80 kg/mm^2). Apparently this is due both to the solubility of carbon in the α-zirconium and to the presence of a martensite $\beta \to \alpha'$ transformation.

The phase diagram of the Zr–C system plotted from the foregoing experimental data is shown in Fig. 11. The melting points of the alloys and their phase composition at various temperatures are indicated on the same graph.

According to the results obtained, the solidus line consists of four parts. In the first part, from 0 to 1 at.% carbon, crystallization of the β phase, constituting a solid solution of carbon in β-zirconium, is completed. On the horizontal part, from 1 to 31 at.% carbon, there is a eutectic reaction $L \rightleftharpoons ZrC + \beta$ at 1805°C. The composition of the liquid taking part in this reaction could not be determined with accuracy. In the third part, the carbide phase, constituting a solid solution of zirconium in the carbide, crystallizes completely. The limiting solubility of zirconium in the carbide at the temperature of the eutectic transformation is 19 at.% zirconium. On the last horizontal section, at 2910°C, the ZrC + C mixture crystallizes. The eutectic point lies at approximately 61 at.% carbon.

There is a single intermediate phase, ZrC, in the Zr–C system; this melts with an open maximum at a temperature near 3500°C.

According to the phase diagram, the introduction of carbon raises the temperature of the $\alpha \to \beta$ transformation of zirconium at the peritectoid composition to 870°C. The solubility of carbon in β-zirconium is negligibly small, but in α-zirconium it reaches a value close to 0.4 wt.%. Judging from the data of the x-ray structural analysis, zirconium does not dissolve in carbon.

Literature Cited

1. Ruff, O., and Wallstein, R. Z. Tech. Phys., 128:100 (1923).
2. Samsonov, G. V., and Umanskii, Ya. F. Hard Compounds of Refractory Metals, Moscow, Metallurgizdat, 1957, p. 105.
3. Becker, K. Hochschmelzende Hortstoffe und ihre technische Anwendung, Berlin (1933).
4. Goldschmidt, U. I. J. Iron Steel Inst., 160:345 (1948).
5. Kieffer, R. Metall, 4(7-8):132 (1950).
6. Rossini, F. D., et al. Natl. Bur. Std. (U.S.), Circ. 500, Ser. 11 (1952).
7. Benesovsky, F., and Rudy, E. Planseeber. Pulvermet., 8:(1960).

8.   Metallurgy of Zirconium [Russian translation], G. A. Meerson and Yu. V. Gagarinskii (eds.), Moscow, IL, 1959, p. 259.

9.   Reports of the United States Atomic-Energy Commission. Nuclear Reactors. Vol. 3. Materials for Nuclear Reactors [Russian translation], Moscow, IL, 1956.

10.   Hansen, M., and Anderko, K. Constitution of Binary Alloys [Russian translation], Moscow, Metallurgizdat, 1962. [English edition: New York, McGraw-Hill, 3rd ed., 1958.]

11.   Godin, Yu. G., et al. In collection: Metallurgy and Metallography of Pure Metals, No. 3, Moscow, Gosatomizdat, 1961, p. 284.

PROPERTIES OF IODIDE-TYPE HAFNIUM
AND ALLOYS IN THE ZIRCONIUM—HAFNIUM SYSTEM

A. I. Evstyukhin, I. P. Barinov, and I. I. Korobkov

Hafnium and zirconium are similar to one another in nature and are obtained from the same raw minerals. The elements are very similar in chemical properties, and this complicates the metallurgical processes of their extraction. Whereas in some cases the complete separation of one metal from another is essential, for example, when using zirconium in atomic technology, in other cases this is not so (use of zirconium in chemical and other branches of technology), since demands are entirely satisfied in this case by the Zr-2-to-2.5% Hf alloys obtained by ordinary treatment of the natural raw material (zircon).

Other alloys in the Zr—Hf system obtained from enriched raw material from factories producing reactor zirconium may also be of interest. There have been several recent papers on various properties of Zr—Hf alloys. The mechanical properties and corrosion resistance of these alloys in water vapor were studied in [1], the interaction of oxides formed on them at high temperatures was considered in [2], and the diffusion of oxygen into zirconium and hafnium was measured in [3, 4]. Information on other physical, mechanical, and chemical properties of Zr—Hf alloys has also been published [5]. The aim of the present investigation was to extend and refine knowledge regarding the properties of hafnium and hafnium—zirconium alloys obtained from Soviet raw materials. The melting points, hardness, and densities of iodide-type hafnium and its zirconium alloys were measured. The kinetics of the oxidation of various Zr—Hf alloys in air and oxygen at 800°C were also studied.

Experimental Part

Iodide Hafnium. Powders of hafnium and its zirconium alloys were obtained by the reaction of oxides containing corresponding quantities of hafnium and zirconium with distilled calcium. The washed and dried powders were converted to compact metals by the iodide method, analogous to that used in the purification of hafnium [6]. In obtaining hafnium and its alloys by the iodide method, it was noted that the raw material had to be more carefully degassed than in the case of producing pure zirconium. The powder had first to be briquetted and converted to chips. The chips had to be degassed at temperatures exceeding 800°C in a vacuum of at least 1×10^{-5} mm Hg. Under other conditions the adsorbed water vapor interacts with carbon impurity in the raw material and forms C_2H_2 and CO, which then dissociate on the hot surface with the production of C, O, and H, the latter not reacting with the hot surface, but again interacting with carbon in the mineral. This may be a reason for the high brittleness of hafnium and hafnium-alloy bars. Table 1 gives the average values of the microhardness of iodide-type hafnium obtained from powder, briquettes, and chippings under identical conditions.

TABLE 1. Average Microhardnesses of
Iodide Hafnium Obtained by the Calcium-
Thermal Method from Metal in Different
Forms

Form of raw material	Micro-hardness, kg/mm²
Powder	244
Briquettes	224
Chips	213

TABLE 2. Chemical Analysis of Calcium-Thermal Powder
and the Iodide Hafnium Obtained from This

Element	Impurity content, wt. %		Element	Impurity content, wt. %	
	original Hf	iodide-type Hf		original Hf	iodide-type Hf
Hf	>99	>99	Fe	—	0.01
Zr	<1	<1	O₂	0.4 (calcium-thermal)	0.014
Al	0.02	0.008			
Ca	0.10	0.02			
Si	0.05	0.01	N₂	0.1 (calcium-thermal)	0.008
Ti	0.09	0.01			
Mn	0.01	<0.005			
Cr	0.003	<0.001			

TABLE 3. Hardness of Iodide Hafnium Subjected to Various
Kinds of Treatment

Iodide bar		Iodide Hf re-melted in arc furnace		Iodide Hf re-melted in electron-beam furnace		Rolled iodide Hf annealed at 850 to 900°C in vacuum (1.5 h)	
R_b	H_B	R_b	H_B	R_b	H_B	R_b	H_B
68	121	80	150	60	107	75	137
63	112	88	176	62	110	73	132
65	116	83	153	57	103	74	135

The brittleness of hafnium and hafnium-alloy bars is also partly due to high iron content in the original material (over 0.25 wt.%); this may fall into the hafnium and be transferred to the hot surface in the course of refining, leading to "red shortness" (weakening of crystal cohesion).

The chemical analysis of the original calcium-thermal powder and that of the resultant bars of iodide-type hafnium are recorded in Table 2 and the corresponding hardnesses in Table 3.

After remelting in an arc furnace with a nonconsumable electrode, the hardness of the iodide hafnium rose on an average by 45 kg/mm²; after remelting in an electron-beam furnace it even fell slightly (from 115 to 107 kg/mm²). After rolling and vacuum annealing, the hardness rose by 20 kg/mm² (to 135 kg/mm²).

Published data on the hardness of iodide hafnium are extremely contradictory (Table 4).

TABLE 4. Hardness of Iodide Hafnium According to Various Authors

Author	Hardness of iodide-type hafnium
De-Boer and Fast, 1930 [7]	"Iodide Hf has a very high ductility, possibly more than Zr"
Carlson, 1957 [8]	69 (scale R_a, for Zr 45); Hf purity 99.8%, large C, O, N content
Litton, 1951 [9]	78 (scale R_b); 206 (scale H_V)
Adenstedt, 1952 [10]	43 (scale R_a); 78 (scale R_b)
Duwez, 1952 [11]	150 (scale H_V)
Gudwin, 1955 [12]	88 (scale R_b, rolled and annealed at 927°C); 160 to 180 (scale H_B, remelted in arc furnace)
Evstyukhin et al., 1964 [13]	65 (scale R_b, rods); 84 (scale R_b, remelted in arc); 60 (scale R_b, remelted in electron-beam furnace); 74 (scale R_b, rolled and annealed at 800 to 900°C)

TABLE 5. Chemical Analysis of Zr−Hf Alloys Obtained
by the Iodide Method

Hf content, wt. %	Impurity content, wt.%									
	N	O	Fe	Si	Al	Ti	Mo	Mn	Ni	Cr
30	0.002	0.02	0.08	0.035	0.01	0.02	0.02	0.01	0.01	0.06
50	0.003	0.02	0.04	0.045	0.028	0.06	0.06	0.01	0.008	0.008
80	0.004	0.02	0.03	0.04	0.035	0.03	0.03	0.01	0.01	0.01
95	0.004	0.02	0.014	0.03	0.03	0.02	0.02	0.01	0.016	0.01
99*	0.002	0.014	0.005	0.01	0.008	0.01	—	0.01	—	<0.005

* Alloy was twice refined.

Such a large spread in the hardness values of iodide hafnium obtained by different authors may be due to different concentrations of impurities, especially oxygen and carbon. In order to check the effect of oxygen on the hardness of hafnium, the authors made some experiments, introducing oxygen deliberately and then measuring the hardness. It was found that iodide hafnium was much more sensitive to oxygen than zirconium (Figs. 1 and 2). The curves relating hardness and microhardness to oxygen content are much sharper for hafnium than for zirconium. The rise in hardness with increasing oxygen content, however, only occurs up to 600°C for hafnium. At higher temperatures the hot-hardness curves level out and then move smoothly downward (Fig. 3).

Zirconium-Hafnium Alloys. Zirconium-hafnium alloys were obtained by the iodide method, by refining raw material of the given composition. The composition and impurity content of such alloys are shown in Table 5.

The resultant alloys were remelted in a MIFI-9-3 arc furnace and studied in the cast, remelted, and rolled state. The density of the alloys were determined pycnometrically for the cast samples. It was found that the density of the alloys rose with increasing hafnium content (Fig. 4).

The melting point of the alloys was also determined on cast samples in the form of rectangular bars; holes 0.8 mm in diameter and 5 to 6 mm deep were drilled in these in order to imitate an absolutely black body. The sample was clamped between two water-cooled copper

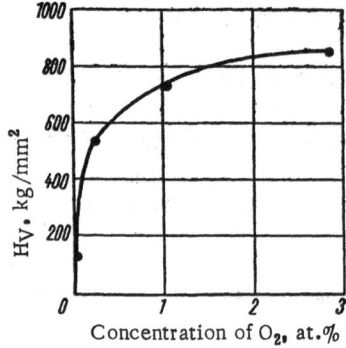

Fig. 1. Microhardness of iodide hafnium as a function of oxygen content.

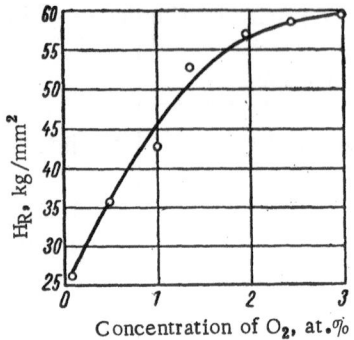

Fig. 2. Hardness of iodide zirconium as a function of oxygen content.

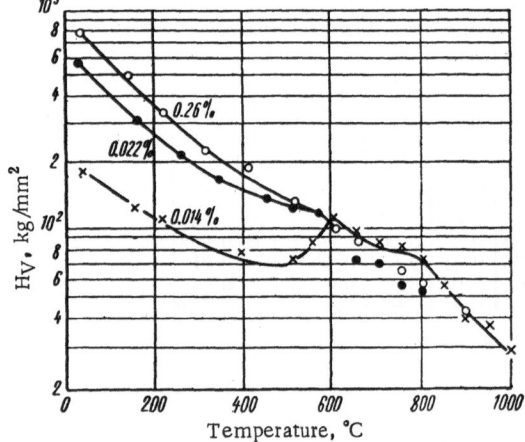

Fig. 3. Hot microhardness of hafnium with various oxygen concentrations (shown on the curves) as a function of temperature.

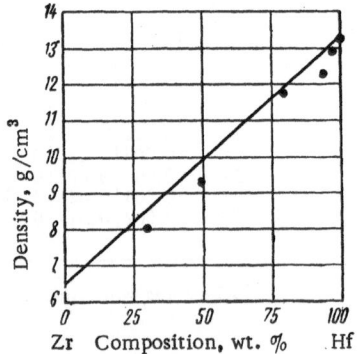

Fig. 4. Density of Zr−Hf alloys as a function of composition.

electrodes carrying an electric current, and heated in an atmosphere of purified helium. The temperature was measured with an OP-48M optical pyrometer, the accuracy of which in the temperature range 900 to 2000°C was 0.2%. The temperature was identified by the appearance of a liquid drop in a hole in the sample.

On raising the hafnium content in the zirconium to 50 wt.%, the melting point of the alloys rose only slightly (80° in all), but, on further increasing the hafnium content, the liquidus of the system rose sharply, especially for hafnium contents above 90 wt.% (Fig. 5). This kind of behavior in the liquidus is characteristic of systems with full mutual solubility of the components and a considerable difference between the melting points.

The microhardness of iodide rods of the alloys and samples remelted from these in the arc furnace was measured as representative of the mechanical properties. The microhardness was determined with a PMT-3 hardness tester, using a 50-g load. It was found that the alloys became harder on raising the hafnium content, the maximum effect being observed at 85 to 90 wt.% Hf (Fig. 6).

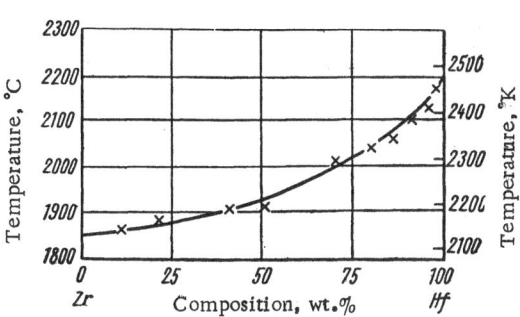

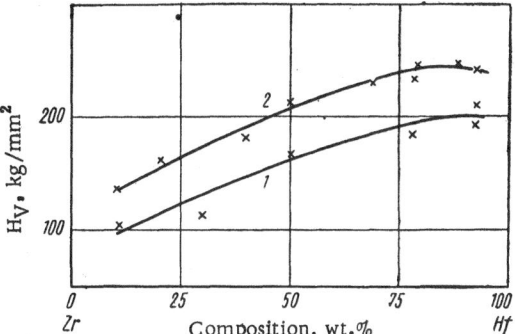

Fig. 5. Liquidus line of the Zr—Hf system.

Fig. 6. Microhardness of Zr—Hf alloys as a function of composition: 1) Iodide rods; 2) rods remelted in the electric-arc furnace.

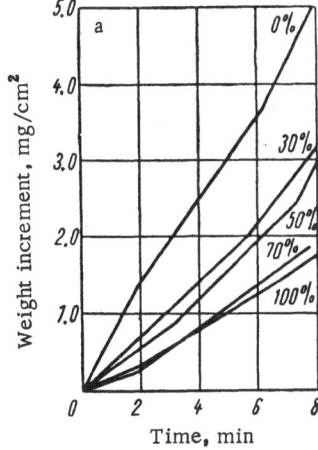

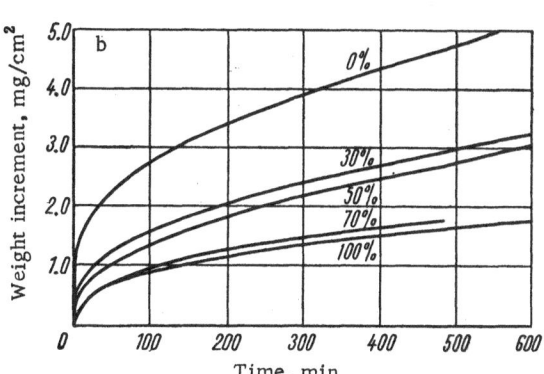

Fig. 7. Kinetics of the oxidation of iodide Zr—Hf alloys in oxygen (P = 150 mm Hg) at 800°C in the initial (a) and steady (b) periods (curves indicate Hf content).

The kinetics of oxidation were studied for Zr—Hf alloys obtained by the iodide method and also for alloys remelted in the arc furnace from pure rods of the iodide metals. In the first case the samples for the corrosion tests were prepared from 4-mm diameter iodide rods by cold forging into strips 1 to 1.5 mm thick and cutting into plates 15 mm long. In the second case the samples for corrosion testing were cut from the ingots in the form of disks and cold-worked to a thickness of 1.5 mm. Samples 9 by 9 mm in size were then cut from the resultant plates. All the samples were ground and vacuum annealed (1×10^{-6} mm Hg, 840°C, 1 h) before testing.

The samples were oxidized both in air and in an oxygen atmosphere (P_{O_2} = 150 mm Hg) by the method described in [14]. In studying the kinetics of oxidation for Zr—Hf alloys it was found that the heat resistance of zirconium in oxygen was continuously improved as its hafnium content increased (Fig. 7a and b). On the other hand, alloying hafnium with zirconium reduced the heat resistance, only very slightly up to 30 wt.% Zr, and then much more rapidly. Similar laws held for the oxidation of Zr—Hf alloys remelted from the pure components in air (Fig. 8a and b). It follows from the kinetic oxidation curves that the alloys have two stages

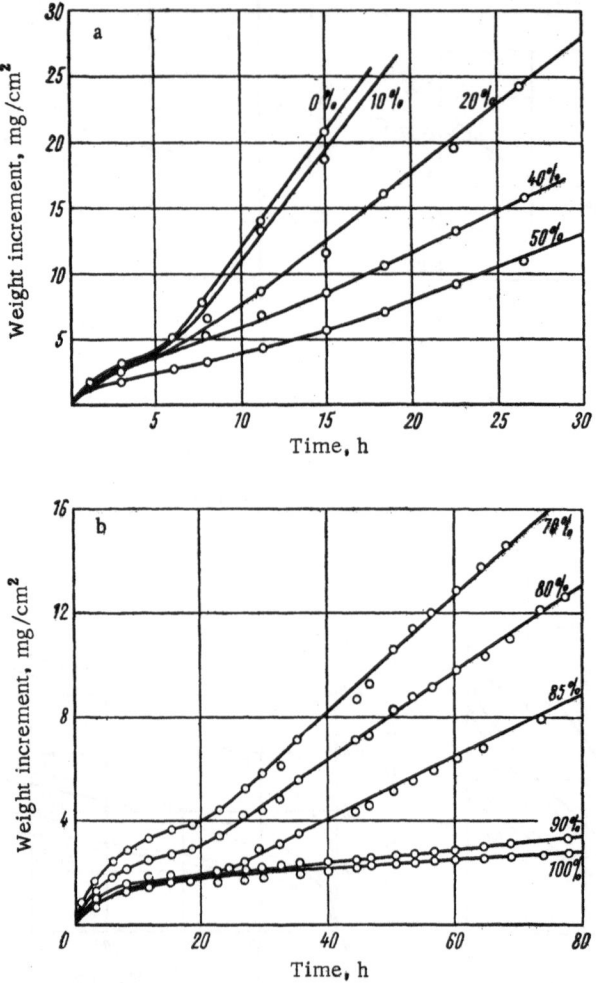

Fig. 8. Kinetics of the oxidation of Zr−Hf alloys
(remelted from the pure components) in air at 800°C
(hafnium content shown on the curves).

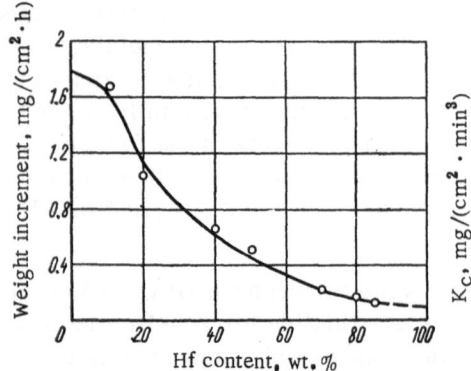

Fig. 9. Rate of oxidation of Zr−Hf
alloys in the steady state as a function of composition.

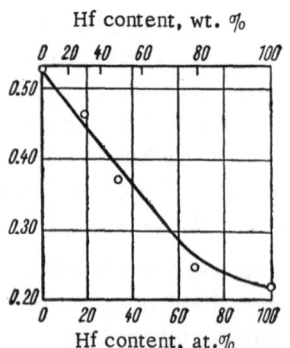

Fig. 10. Reaction-rate constant for the oxidation of Zr−Hf
alloy in the initial period as a
function of composition.

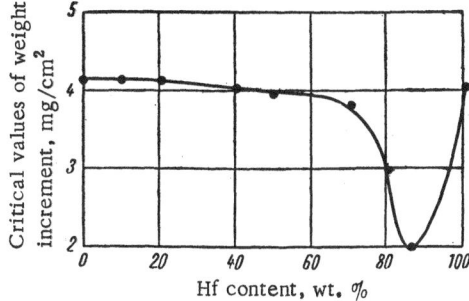

Fig. 11. Critical values of weight increment as functions of the composition of the Zr−Hf alloys corresponding to the bend in the kinetic curves at which the oxidation rate begins to increase.

of oxidation similar to those of the pure zirconium and hafnium components. The first (initial) stage, the duration of which depends on the composition of the alloy, is characterized by the growth of a continuous dark film on the surface and a fall in the oxidation rate with time (on a cubic or parabolic law). The beginning of the second (steady) stage coincides in time with the appearance of a white porous scale on the sample surface (first in the form of individual points, the dimensions of which increase with time). This stage is characterized by an increase in weight resulting from the cracking of the protective film and the growth beneath this of a new oxide underlayer. After a little while this process is stabilized and the oxidation law becomes linear.

Figure 9 shows the oxidation rate of Zr−Hf alloys in the steady oxidation process as a function of their composition. This relationship is characterized by a smooth curve analogous to that of the oxidation reaction-rate constant in the initial stage (Fig. 10).

Figure 11 shows a relation between the critical weight increment and the composition of alloys corresponding to the bend in the kinetic curves at which the oxidation rate begins rising. We notice a minimum on the critical weight-increment curve. This corresponds to the fact that in samples containing 80 to 90 wt.% Hf, the protective films formed by the oxidation thicken less before breaking than in other alloys.

Conclusions

1. The hardness and microhardness of iodide-type hafnium have been determined in various states and the effects of oxygen and other impurities examined.

2. The liquidus line of the Zr−Hf system has been determined experimentally for alloys prepared by the iodide method.

3. The variation in the density and hardness of Zr−Hf alloys has been determined in terms of composition.

4. The kinetics of the oxidation of Zr−Hf alloys in air and oxygen at 800°C have been studied, using iodide samples of the alloys and other samples remelted from the pure iodide-type metals.

Literature Cited

1. Grebennikov, R. V., and Shamashov, F. P. Corrosion of Reactor Materials, Vol. 2, Moscow, Izd. MAGATÉ, 1962, pp. 149-158.
2. Curtis, C. E., et al. J. Am. Ceram. Soc., 37:458 (1954).
3. Pemsler, J. P. J. Electrochem. Soc., 12:12 (1959).
4. Orr, R. L. J. Am. Chem. Soc., 75:1231 (1953).
5. Ray, V. E. Production of Control Rods for Nuclear Reactors [Russian translation], I. S. Golovin (ed.), Moscow, Atomizdat, 1965.
6. Emel'yanov, V. S., et. al. In collection: Metallurgy and Metallography of Pure Metals, No. 1, Moscow, Izd. MIFI, 1959, p. 63.
7. De-Boer, J. H., and Fast, J. D. Z. Anorg. Allgem. Chem., 187:193 (1930).
8. Carlson, O. N., Schmidt, F. A., and Wilhelm, H. A. J. Electrochem. Soc., 104:51 (1957).

9.   Litton, F. B. J. Electrochem. Soc., 98:488 (1951).
10. Adenstedt, H. K. Trans. Am. Soc. Metals, 44:949 (1952).
11. Duwez, P., et al. Phys. Rev., 85:989 (1952).
12. Gudwin, J. G., and Hurford, W. J. J. Metals, 7:1162 (1955).
13. Evstyukhin, A. I., et al. This volume, p. 23.
14. Revyakin, B. N., et al. In collection: Metallurgy and Metallography of Pure Metals, No. 3, Moscow, Gosatomizdat, 1961, p. 175.

EFFECT OF OXYGEN IMPURITY ON THE STRUCTURE
AND SUPERCONDUCTING PROPERTIES
OF ZIRCONIUM–NIOBIUM ALLOYS

Yu. F. Bychkov, I. N. Goncharov, and I. S. Khukhareva

Zirconium alloys containing 20 to 35% niobium are of special interest as material for constructing superconducting magnets, since a Zr–25% Nb alloy has the greatest value of upper critical field (H_{C2} = 125 to 150 kOe [1]) in the whole Zr–Nb system, and by due heat and mechanical treatment of these alloys high values of the critical current density in a 30-kOe field (10^5 A/cm^2) can be obtained [2]. It is well known that, in contrast to the upper critical magnetic field, the critical current density is a structure-sensitive property, the value of which is greatly affected by the presence of impurities, the nature and degree of mechanical deformation (working), and the presence and distribution of second-phase inclusions.

In the present investigation, we studied the effects of oxygen impurities on the critical current density J_C and on the decomposition of the β solid solution, which also affected its value.

It is known from [3] that the addition of 0.025 to 0.05 wt.% oxygen to a Zr – 75% Nb alloys leads to a considerable rise in J_C if the alloy is subjected to intermediate annealing for 15 min at 800°C, but produces no change in J_C if the alloy is tested in the cold-worked state. There is no such evidence as regards zirconium-base alloys.

It is also known that impurities of oxygen and nitrogen lead to a rise in H_{C2} and J_C for niobium [4]; at 4.2°K, the H_{C2} for high-purity niobium, with a ratio of the electrical resistance at room temperature to the residual resistance equal to R_{300}/R_{res} = 505, is 2 kOe, while for less pure niobium with R_{300}/R_{res} = 3.1, H_{C2} = 8 kOe, although the corresponding critical temperature T_C falls from 9.1 to 7.7°K.

Niobium contaminated with gas impurities (with R_{300}/R_{res} = 2.45 to 4.67) has a 4-to-10 times higher J_C than the purer material with R_{300}/R_{res} = 100.

We considered the effect of intermediate annealing on J_C for Zr–Nb alloys containing up to 0.25 wt.% oxygen impurities. It is desirable to study the effects of intermediate annealing on the J_C of these alloys because, on the one hand, intermediate annealing at 400 to 570°C sharply increases J_C in Zr-20-to-35% Nb alloys not containing oxygen [2], and, on the other hand, oxygen impurities have a considerable effect on processes underlying the decomposition of the β solid solution of Zr–Nb, changing the form of the phase diagram.

The effect of oxygen impurities in the original metals on the form of the Zr–Nb phase diagram was analyzed for the first time in [5]. The authors of [5] plotted two diagrams, one

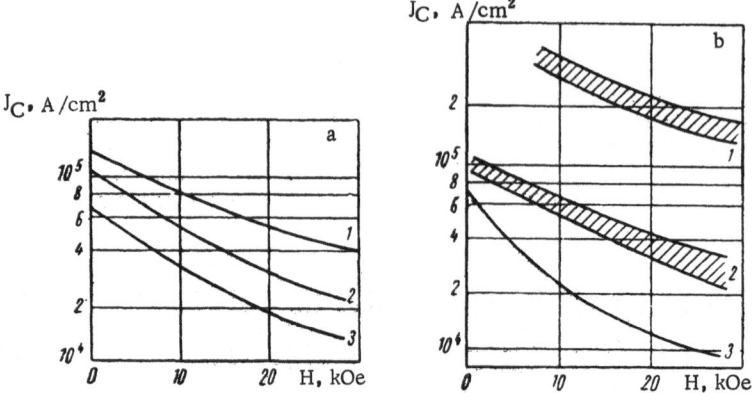

Fig. 1. Effect of oxygen on the critical current density for Zr−Nb alloys cold-rolled from 1 to 0.05 mm after homogenization at 1300°C (a) and intermediate annealing at 700°C (b): 1) 33% Nb + 0.25% O (reduction 17:1); 2) 33% Nb + 0.1% O (reduction 17:1); 3) 33% Nb, without oxygen (reduction 200:1) or with 0.04% O (reduction 17:1).

using zirconium sponge base for preparing the alloys, and the other using the purer iodide-type zirconium. The zirconium sponge and its alloys with 2 and 20% niobium contained 0.015 wt.% oxygen, whereas the alloys formed from the iodide zirconium contained four times less (0.004 wt.%). The phase diagrams plotted were regarded as sections of the ternary Zr−Nb−O system with constant oxygen content. The change from iodide-type to zirconium sponge broadened the interval of the $\beta \rightarrow \alpha + \beta'$ transformations for a niobium content of 18 wt.% (monotectoid point). The lower boundary of the region of $\beta \rightarrow \alpha + \beta'$ transformations lay at approximately 570 to 590°C. The great differences in the position of the monotectoid points on the phase diagrams (from 12 to 20% niobium in composition and from 560 to 800°C in temperature) obtained by different authors was explained in [5] as being due to different impurity contents in the original metals.

The effect of oxygen impurity on the form of the Zr−Nb phase diagram was also studied in [6]. According to this paper, the addition of 0.25 wt.% oxygen leads to the phase separation of the β solid solutions containing 75% niobium after annealing at 800°C for 7 h, giving two β solid solutions of different concentration, whereas alloys free from added oxygen show no phase separation of the β phase, provided that a 0.1-mm layer is removed from the surface before x-ray phase analysis. The existence of phase separation on the surface of annealed strip was attributed by Berghout [6] to contamination of the surface of the microsection with oxygen and nitrogen during the annealing process.

We ourselves studied a Zr−33 wt.% Nb alloy into which various quantities of oxygen had been introduced. The alloy was prepared in an electron-beam furnace from iodide zirconium and niobium pieces. After remelting, the alloys contained 0.02 to 0.04 wt.% oxygen. Bars some 55 mm in diameter were forged at 800 to 900°C in air and plate 15 mm thick was obtained. After removing the scale, the material was cold-rolled from 10 to 1 mm. Plates 1 × 20 × 50 mm in size were cut from the resultant strip; these were oxygen-saturated in the apparatus described in [7]. Oxygen of commercial purity was used. The samples, suspended in a quartz flask, were vacuum-degassed at 800°C, weighed, and again placed in the flask. After introducing sufficient oxygen to produce the required concentration in the alloy into the flask, the sample was heated to 900 to 930°C. The absorption of oxygen was checked by mercury and oil manometers to an accuracy of 1% of the oxygen introduced. In order to secure a uniform distribution of the oxygen over the cross section of the strip, homogenization was carried out

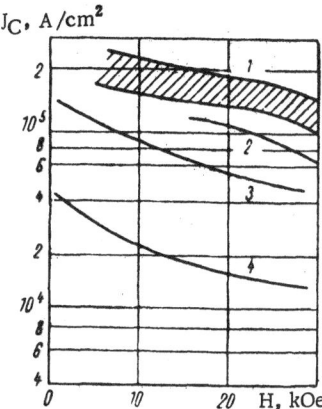

Fig. 2. Effect of intermediate annealing at 500, 570, and 700°C and subsequent rolling from 0.5 to 0.05 mm on the J_C of Zr—33% Nb — 0.25% O: 1) Annealing at 500 and 570°C; 2) annealing at 700°C; 3) without annealing; 4) without oxygen and without annealing.

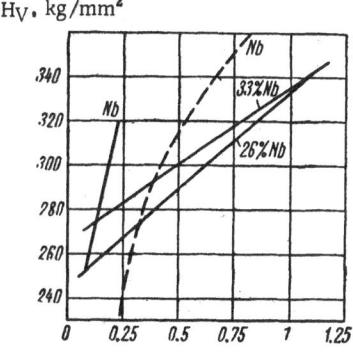

Fig. 3. Hardness H_V of alloys containing 26 and 33% niobium and hardness H_B of niobium as functions of oxygen content.

in a TVV-4 furnace for 1.5 h at 1300°C in a vacuum of 10^{-3} to 10^{-4} mm Hg. During homogenization, additional contamination with oxygen occurred. By weighing the samples on an analytical balance, the additional quantity of oxygen absorbed was determined. It is well known that Zr—Nb alloys interact mainly with oxygen and not nitrogen in air at 1300°C.

The oxygen-saturated and homogenized strip was cold-rolled to 0.5 mm and subjected to intermediate annealing at 500, 570, and 700°C for 1 h and then cold-rolled to 0.05 mm. In strips of this kind 0.6 mm broad, the critical current density in a magnetic field parallel to the rolling plane was measured. For the sake of comparison, samples cold-rolled directly from 1 to 0.05 mm immediately after homogenization (without any intermediate annealing) were also tested.

Results of measuring J_C in such alloys are shown in Fig. 1a. Addition of 0.04 wt.% oxygen hardly affected the value of J_C; the relation between J_C and H was the same as in the case of an alloy subjected to cold work after quenching from the β region (from a temperature exceeding 800°C) without adding oxygen (see Fig. 1a, curve 3). In the alloy which was simply subjected to cold work after homogenization, the critical current density more than doubled on raising the oxygen content from 0.1 to 0.25 wt.%, reaching $4 \cdot 10^4$ A/cm^2 for a field of 27 kOe (Fig. 1a).

Oxygen raises the critical current density still more sharply if the alloys are subjected to intermediate annealing and subsequent cold-rolling. The values of J_C after intermediate annealing at 500 to 700°C are shown in Figs. 1b and 2. The figures show that for a high oxygen content the value of J_C rises to $1.5 \cdot 10^5$ A/cm^2 at a field of 27 kOe.

The increase of oxygen concentration in the alloys is accompanied by a considerable rise in hardness, though to a less marked extent than in the case of unalloyed niobium or niobium-base alloys; the data for niobium are taken from [8] (Fig. 3). Apparently alloys containing 20 to 35 wt.% niobium permit the introduction of greater quantities of oxygen than niobium-base alloys, owing to the smaller effect of oxygen on their ductility. After annealing at 700°C, alloys containing 33 wt.% Nb and 0.25 wt.% O were hardly softened at all, while the hardness of the alloy containing 26 wt.% Nb and 0.04 wt.% O fell by 20 units.

Metallographic study of Zr—33% Nb—0.18% O after homogenization at 1300°C showed that this alloy constituted a solid solution with a small quantity of second-phase inclusions, the dimensions of which were approximately 3×10^{-4} cm (Fig. 4). After annealing at 500 to 700°C, all the oxygen-containing samples studied [26% Nb—0.04 (0.18)% O and 33% Nb—0.1 (0.18 or 0.25)% O] were two-phased. The development the second phase after annealing at 700°C was primarily due to the phase separation of the β solid solution in the presence of oxygen during the annealing period.

Amount of α Phase in Zr−Nb Alloys Containing Oxygen after Annealing for 1 h at 570°C

Part analyzed	Alloy composition							
	26% Nb	26% Nb − 0.18% O	26% Nb − 0.5% O	26% Nb − 1% O	33% Nb − 0.1% O	33% Nb − 0.18% O	33% Nb − 0.25% O	33% Nb − 1% O
Surface	Much	Much	Much	Much	Much	Much	Much	Much
At a depth of:								
0.1 mm	None	Very little	Very little	—	Very little	Very little	Very little	—
0.2 mm	—	Same	Same	—	None	Same	Same	—
0.3 mm	—	Same	Same	—	—	Same	Same	—
0.4 mm	—	Same	—	Much	—	—	—	Much

Fig. 4. Microstructure of Zr− 33% Nb− 0.18% O after homogenization at 1300°C.

As indicated earlier [2,6], the structure on the surface of the alloy containing 20 wt.% Nb after annealing at 570°C (i.e., below the monotectoid temperature) differs considerably from that inside the sample. X-ray analysis reveals three phases on the surface: the original β phase, the α phase, and a β' phase containing 85 wt.% Nb; inside the sample, at a depth of 0.05 mm or over, only the original β phase occurs. The surface decomposition may be accelerated either by the more severe deformation given to the surface or by its contamination with gases during the annealing period.

The effect of oxygen on the decomposition of the β solid solution within the samples was determined after annealing at 570°C (below the monotectoid temperature) for 1 h. The occurrence of decomposition in the β solid solution was indicated by the development of the α phase (see table). In the original alloy containing 26 wt.% Nb and not saturated with oxygen, the α phase was present on the surface together with the original β phase but there was no α phase at a depth of 0.1 mm into the sample. In alloys containing 0.18 to 0.5 wt.% oxygen, the surface structure also differed from that inside the sample, where the α phase (solid solution of O in Zr) was only present in minute quantities. This phase was not formed as a result of the break-up of the β phase, but as a result of the α phase by oxygen [11]. The α phase inside the samples was formed because the oxygen tended to combine principally with the zirconium, forming a solid solution Zr−O. Since this phase was also present in samples not subjected to intermediate annealing, it was presumably to be associated with the increase in the J_C of alloys containing oxygen. In alloys containing 1 wt.% oxygen, the amount of α phase was unvarying and quite appreciable.

Thus the introduction of small traces of oxygen into zirconium-base alloys enables J_C to be considerably raised, especially if intermediate annealing followed by cold-rolling is employed.

Oxygen impurities lead to a rise in the hardness and cause the separation of the β solid solution into two solid solutions of different concentrations. The rise in J_C in the presence of oxygen inclusions in these alloys may be explained on the Anderson model [9, 10] by the fact that the presence of dispersed inclusions with poor superconducting properties (precipitates of α-zirconium, with $J_C = 0.6°K$, in particular) in the superconducting matrix has a stabilizing influence on the lines of magnetic flux.

These results constitute further evidence of the fact that J_C varies with structure, but further study is required in order to decide which elements of structure have the greatest effect on J_C.

Literature Cited

1. Berliucourt, T. G., and Hake, R. R. Phys. Rev. Let., 9:293 (1962).
2. Bychkov, Yu. F., et al. Pribory i Tekhn. Eksperim., No. 3:170 (1964).
3. Betterton, J. O., et al. Superconductors, New York, Interscience, 1962, p. 61.
4. Rosenblum, E. S., and Autler, S. H. Rev. Mod. Phys., 36:77 (1964).
5. Richter, J. Less-Common Metals, 4:252 (1962).
6. Berghout, C. W. Phys. Let., 1:292 (1962).
7. Borkov, N. V. In collection: Metallurgy and Metallography of Pure Metals, No. 2, Moscow, Atomizdat, 1960, p. 148.
8. Prokoshkin, D. A., and Vasil'eva, E. V. Niobium Alloys, Moscow, Nauka, 1964.
9. Anderson, P. W. Phys. Rev. Let., 9:309 (1962).
10. Anderson, P. W., and Kim, Y. B. Rev. Mod. Phys., 36:39 (1964).
11. Marcotte, V. C., et al. J. Less-Common Metals, 7:373 (1964).

STUDY OF ALLOYS IN THE NIOBIUM—ZINC SYSTEM

A. I. Evstyukhin, Yu. G. Godin, and V. B. Yakovleva

A study of niobium–zinc alloys is of special interest in view of their high resistance to oxidation at high temperatures. Such alloys might well find application in a number of branches of new technology. There is nevertheless very little information regarding them in the literature. This is apparently because of the complexity of preparing the alloys (due to the volatility of the zinc) and the difficulty of treating and examining them. It is no doubt for the same reason that the phase diagram and phase transformations of the Nb—Zn system have not been studied. The only published information [1, 2] is that a number of compounds exist in this system.

Our present problem was to obtain information on various properties of niobium–zinc alloys. It is also of interest to determine the reasons for the heat resistance of the alloys and to determine the properties of the oxides formed on them.

Method of Investigation

Preparation of the Alloys. In order to prepare alloys of the Nb—Zn systems, we used granulated, chemically pure zinc and niobium pieces of 99.8% purity. In selecting a method for preparing the alloys, we took account of the unusually high vapor pressure of zinc at temperatures close to the melting point of niobium (Table 1).

We tried several methods of preparation before achieving satisfactory results. First we tried to prepare alloys in an arc furnace by melting a charge composed of solid pieces of niobium and zinc, and also by feeding pieces of zinc from a bunker into molten niobium. Both these methods proved unsatisfactory. In the first case the zinc almost entirely evaporated before the niobium melted, and in the second case the zinc falling into the molten niobium caused a considerable rise in temperature and the zinc vapor so formed produced severe splashing of the molten niobium. Better results were obtained by using the following method. The charge was loaded into ceramic crucibles, which were then placed in a Ya1T stainless steel container (Fig. 1). Two crucibles were usually placed in the container. Then the lid of the container was screwed down and the whole was placed in a MIFI-9-3 arc furnace. The furnace and container were evacuated, flushed with argon, and after finally filling with the gas the lid of the container was sealed.

At a given temperature the containers were placed in a Silit furnace and held for a specified time. After holding in the furnace, the container was cut open on a lathe, and the alloys were extracted and subjected to metallographic examination.

TABLE 1. Zinc Vapor Pressure at High
Temperatures [3]

Temperature, °C	Zinc vapor pressure	
	mm Hg	atm
1000	$2 \cdot 10^3$	2,8
1100	$5 \cdot 10^3$	6,6
1200	$7 \cdot 10^3$	9,2
1300	$1 \cdot 10^4$	13
2400	$5 \cdot 10^5$	660

TABLE 2. Conditions for Preparing Nb−Zn Alloys

No. of batch	Temperature, °C	Holding time in furnace, h	Nb content in charge, wt. %	Number of samples
1	1000	2	from 1 to 19	10
2	1000	4	from 1 to 50	18
3	1100	3	from 1 to 40	14
4	1100	6	from 25 to 40	4
5	1200	8	25; 40	2
6	1150	20	from 2 to 40	17

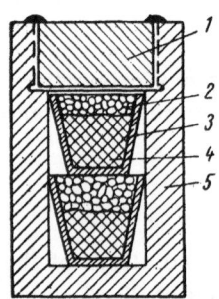

Fig. 1. Stainless-steel
container for producing
Nb−Zn alloys: 1) Lid of
the container; 2) niobium;
3) ceramic crucible; 4)
zinc; 5) container.

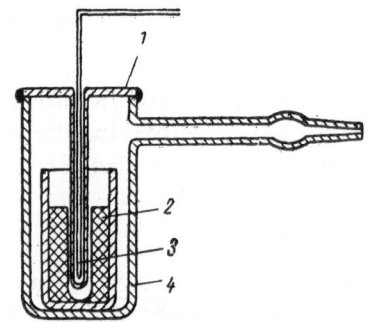

Fig. 2. Quartz capsule and
crucible for thermal analysis :
1) Lid; 2) crucible containing
sample; 3) thermocouple; 4)
quartz capsule.

In many alloys of each batch, a certain amount of nonreacting niobium was found; as the niobium content increased, the quantity of undissolved pieces did likewise. In view of this the holding time was increased to 20 h, the furnace temperature being kept constant at 1150°C (Table 2). Seventeen alloys were obtained under these conditions. All were homogeneous, without inclusions of metallic niobium. The charge composition of the alloys lay between 2 and 40 wt.% Nb and differed considerably from chemical-analysis data (from 4.9 to 52.7 wt.% Nb). Alloys with up to 20 wt.% Nb were compact; those with more niobium were porous.

In order to obtain alloys with a high niobium content, we made several ingot meltings in the MIFI-9-3 arc furnace. The original ingots contained some 30 to 40 wt.% niobium, but as a result of the intensive sublimation of zinc in the melting process the final proportion of niobium in the alloys was much greater, reaching 90 to 100%. It was quite impossible to control

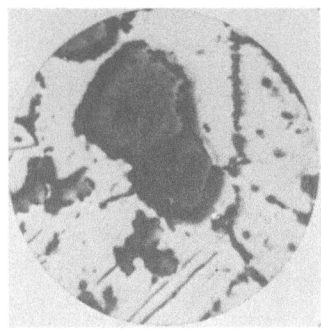

Fig. 3. Structure of an alloy with a small niobium content (× 700).

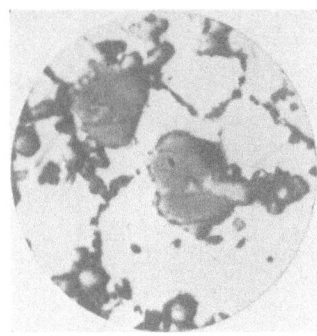

Fig. 4. Structure of an alloy with a 9.88 wt.% content of Nb (× 700).

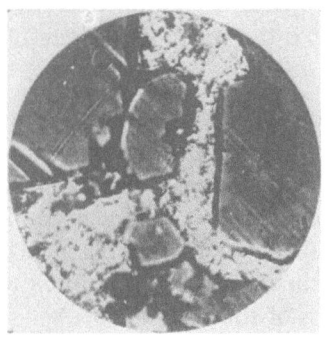

Fig. 5. Structure of an alloy containing 28.4 wt.% Nb (× 1200).

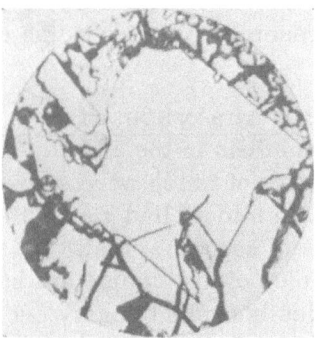

Fig. 6. Structure of an alloy containing 35.8 wt.% Nb (× 1200).

the composition of the alloys in the course of melting. Owing to the condensation of zinc vapor on the observation window it was also difficult to observe the process, especially in the final stages.

The alloys prepared by fusion in containers and remelting in the arc furnace were cut or broken into individual pieces, which were then studied metallographically in order to reveal any undissolved pieces of niobium.

Metallographic Analysis of the Alloys. The preparation of microsections was carried out in the usual way. Since the alloy samples were brittle, they crumbled severely on treatment with emery paper. In order to prevent this, the final pieces of emery paper were rubbed with wax or paraffin. The structure of the alloys was revealed by etching in an aqueous solution of hydrofluoric and nitric acids.

X-Ray Analysis of the Alloys. X-ray diffraction photographs of the alloys were taken in a cylindrical RKU-86 camera, using CuKα radiation. The samples were prepared in the form of cylinders 0.6 mm in diameter by filling a Saponlac capillary with powder sieved through a 150 mesh. A sharp-focus tube was used at a voltage of 40 KV and current 2 ma; the exposure time was 4 h.

Hardness Measurement. The microhardness of the alloys was determined with a PMT-3 hardness tester, using loads of 20 and 50 g. The macrohardness of some alloys was

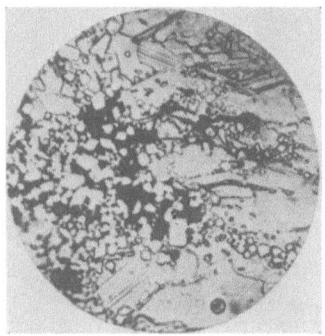

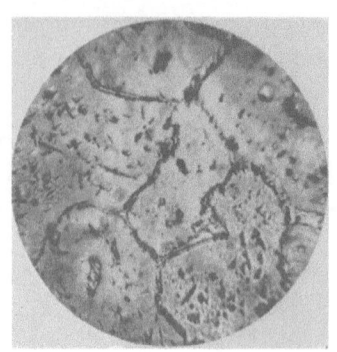

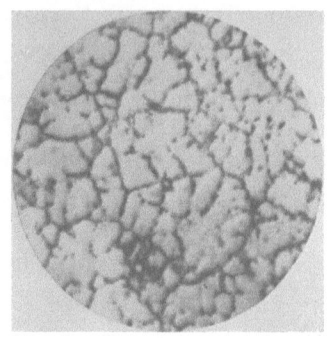

Fig. 7. Structure of an alloy containing 55.57 wt.%
Nb (× 900).

Fig. 8. Structure of an alloy containing 97 wt.% Nb
(× 900).

Fig. 9. Structure of an alloy containing 99.29 wt.%
Nb and 0.48 wt.% Zn (× 270).

measured on a TP tester (diamond tip) with various loads; owing to the brittleness of the samples and the presence of pores, it was impossible to measure the macrohardness of most of the
samples.

Thermal Analysis of the Alloys. Thermograms were plotted in order to study
phase transformations in the alloys. The samples were heated in small argon-filled quartz capsules, into the lids of which were sealed two thin quartz capillaries with dead ends. One capillary was let down into a blind hole drilled in the sample and the other into a hole drilled in a
standard (piece of copper). The lid was sealed to the capsule after the capillaries had been
placed in the holes of the sample and standard. Platinum-rhodium thermocouples were placed
in the capillaries in order to carry out the thermal analysis of the alloys.

Evacuation of the capsules and subsequent filling of these with argon were effected via a
side tube, which was sealed off after admitting the gas. The amount of argon introduced into
the capsule was calculated in such a way that the gas pressure in the capsule should only slightly exceed atmospheric at the maximum temperature. The presence of an inert atmosphere in
the capsule and the maintenance of a uniform temperature over its entire volume reduced the
loss of zinc from the alloys by sublimation in the course of taking the thermograms. In order
to ensure the required thermal inertia in the course of thermal analysis, the capsule was placed
in a thick-walled cylinder made of stainless steel. The equipment for thermal analysis is shown
in Fig. 2.

Constant heating and cooling rates of 6 to 8 deg/min were maintained by means of an autotransformer. The heating and cooling curves were determined on a Kurnakov pyrometer with
automatic recording.

Results

Phase Analysis. Alloys containing small quantities of niobium show two structural components (Figs. 3 and 4). The base of these alloys is zinc, on the light background of
which formations of a niobium-impoverished intermediate phase ($NbZn_3$) can be seen.

As the niobium content of the alloys rises, the amount of $NbZn_3$ does likewise, and the
amount of zinc diminishes.

In an alloy containing 28.4 wt.% niobium, the main field of the section is occupied by
$NbZn_3$ (Fig. 5), between the grains of which zinc crystals may be seen. Increasing the niobium
content in the alloys above 28.4 wt.% causes the zinc to vanish. Figure 6 shows the structure
of an alloy containing 35.8 wt.% niobium; this consists of large crystals of the intermediate

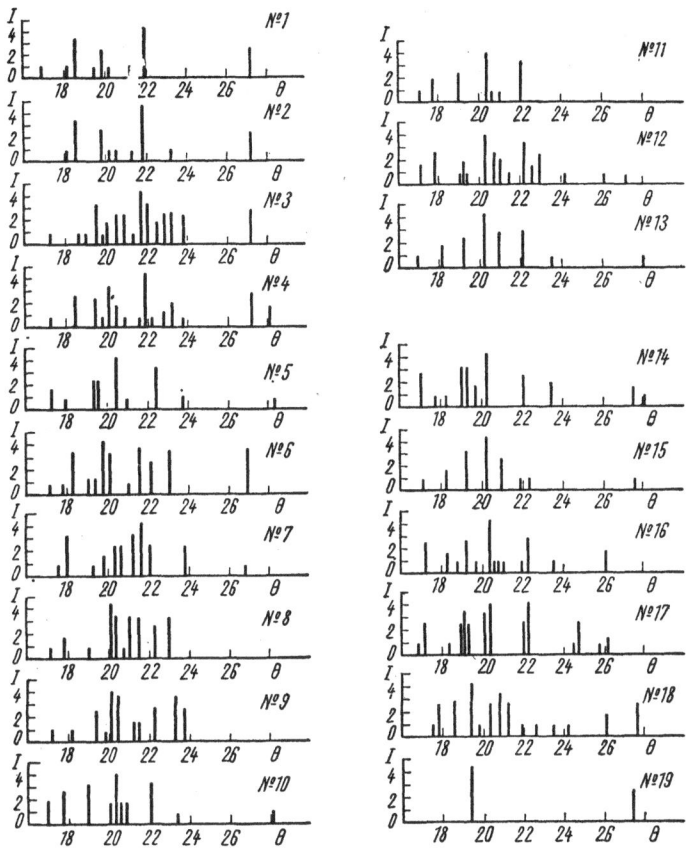

Fig. 10. Schematic representation of x-ray diffraction
photographs (alloys No. 1 to 19).

phase Nb_2Zn_3 and small crystals of the intermetallic compound $NbZn_3$. On further raising the niobium content, the amount of phase Nb_2Zn_3 increases, and in an alloy containing 39.4 wt.% niobium it occupies almost the whole field of the microsection. An alloy containing 52 wt.% niobium consists mainly of homogeneous grains, inside which small inclusions may be seen. We may suppose that the composition of this alloy is close to that of the intermetallic compound Nb_2Zn_3. The small precipitates seen in this alloy may consist of the phase $NbZn$. The structure of an alloy containing a large proportion of niobium (55.57 wt.%) is two-phased (Fig. 7). The main field of the section for this alloy is occupied by grains of Nb_2Zn_3, on the background of which small $NbZn$ crystal may be seen. An alloy containing 59.39 wt.% Nb consists mainly of $NbZn$ grains and a small quantity of Nb_2Zn_3.

The structures of all alloys containing more than 60 wt.% niobium show metallic-niobium crystals. In the alloy containing 74.4 wt.% Nb, the $NbZn$ phase is arranged in the form of a lattice along the boundaries of the niobium grains, which constitute the greater part of the alloy. Further increasing the amount of niobium in the alloys leads to a fall in the number of $NbZn$ crystals. In an alloy containing 97 wt.% niobium, the $NbZn$ phase is arranged in the form of a thin lattice along the niobium grain boundaries (Fig. 8). Figure 9 shows the microstructure of an alloy containing 99.29 wt.% niobium and 0.48 wt.% zinc.

X-Ray Analysis. The x-ray analysis data are shown in the form of schematic x-ray diffraction records (Fig. 10). Comparison of the x-ray diffraction pictures of alloys containing 4.98, 9.88, and 28.4 wt.% niobium with that of zinc shows that the series of lines

TABLE 3. Microhardness of Niobium-Zinc Alloys

Composition of alloy, wt. %		Microhardness of phases, kg/mm²				
Zn	Nb	Zn	NbZn₃	Nb₂Zn₃	NbZn	Nb
95.02	4.98	64	302	—	—	—
90.12	9.88	69	302	—	—	—
80.6	19.4	67	307	—	—	—
64.2	35.8	—	—	925	—	—
59.88	40.12	—	—	890	—	—
55.4	44.5	—	—	890	—	—
50	50	—	—	840	—	—
47.3	52.7	—	—	—	460	—
44.5	55.5	—	—	—	429	—
27.5	72.5	—	—	—	—	206
25.6	74.4	—	—	—	—	228
16	84	—	—	—	—	232

TABLE 4. Macrohardness of Niobium-Zinc Alloys

Composition of alloy, wt.%		Microhardness, kg/mm²
Zn	Nb	
95.02	4.98	51
90.12	9.88	62
80.6	19.4	200
16	84	432
3.1	96.9	389

characteristic of zinc weaken as the niobium concentration increases, while the intensity of the lines belonging to the $NbZn_3$ phase becomes greater. The x-ray picture of the alloy containing 28.4 wt.% Nb contains almost all the reflections of intermetallic compound $NbZn_3$.

The x-ray diffraction picture of the alloy containing 35.8 wt.% niobium shows lines of $NbZn_3$ and $NbZn_2$, the intensities of the reflections belonging to these phases being almost equal. The number of lines on the x-ray picture of the alloy containing 52 wt.% niobium is the smallest. The system of reflections on this x-ray picture differs from those corresponding to the $NbZn_3$ and $NbZn_2$ phases. From the metallographic data and considerations of stoichiometry, we may suppose that the x-ray lines in this picture belong to Nb_2Zn_3.

All the lines corresponding to the Nb_2Zn_3 phase occur on the x-ray picture of the alloy containing 55.57 wt.% niobium. In addition to these, lines corresponding to another phase (apparently NbZn) occur. This is confirmed by comparing the x-ray picture with that of an alloy containing 60.3 wt.% niobium.

The x-ray picture of this latter phase has a small number of lines not corresponding to reflections from $NbZn_3$, $NbZn_2$, and Nb_2Zn_3.

Alloys containing more than 60.3 wt.% niobium have two phases: niobium and NbZn. With increasing niobium content, the intensity of the NbZn lines diminishes, and that of the niobium increases. The presence of a series of weak lines on the x-ray photograph of the alloy containing 74.4 wt.% niobium is apparently due to the presence of foreign impurities.

Hardness and Microhardness of the Alloys. The results of the microhardness measurements are shown in Table 3.

We see from Table 3 that the intermetallic compound Nb_2Zn_3 has the greatest hardness, equal to 890 kg/mm², while that of $NbZn_3$ is 302 kg/mm².

Table 4 shows the measured macrohardness values for the alloys.

The macrohardness of other samples could not be measured owing to their great porosity.

Conclusions

1. We have developed a method of alloying volatile metals (low boiling point) with refractory metals.

2. We have proposed a method for the thermal analysis of such alloys.

3. We have confirmed earlier published data regarding the existence of four intermetallic compounds, $NbZn_3$, $NbZn_2$, Nb_2Zn_3, and $NbZn$ in the Nb—Zn system.

4. We have determined the microhardness of these intermetallic compounds.

5. We have demonstrated that the study of phase transformations in Nb—Zn alloys at temperatures above 1000°C requires the installation of special apparatus enabling the work to be carried out at high pressures.

Literature Cited

1. Acta Cryst., 13:743 (1960).
2. Seeboed, R. E., and Birus, L. S. Naval Research Lab., Washington, D. C., Sept. 19, 1960, p. 13.
3. Reports of the United States Atomic-Energy Commission. Nuclear Reactors. Vol. 3. Materials for Nuclear Reactors [Russian translation], Moscow, IL, 1956, p. 384.

CRYSTALLIZATION FRONT IN
ELECTRIC-ARC ZONE MELTING

I. V. Milov, D. M. Skorov, and V. V. Nikishanov

For low concentrations of impurities, the efficiency of the zone method of purification is considerably affected by disruption of equilibrium near the crystallization front. The reason for this is the nonuniformity of motion at the crystallization front arising from nonuniform motion of the zone or the sample itself, instability in the heater power, inconstancy in the sample cross section, and inhomogeneity of composition along the length of the sample. The main cause of nonequilibrium crystallization conditions, however, is the supercooling of the melt in advance of the crystallization front, resulting from variation of composition in the molten zone.

The cause of the supercooling lies in the variation of the impurity concentration in the layers of liquid closest to the crystallization front, which is due to the incomplete mixing of impurities in the liquid phase. The change in concentration affects the equilibrium temperature. Figure 1 shows the impurity distribution near the crystallization front for the case in which the partition coefficient of the impurity is smaller than unity. As the crystallization front moves forward, some parts, thrust out a little in advance, find themselves in a metastable region of supercooling. Intensive growth of dendritic crystals ensues, with the velocity of growth exceeding the average velocity of the crystallization front (zone velocity).

The impurity-transfer velocity associated with the motion of the zone may be determined from the equation

$$W = C_i U_0 - D_i \left(\frac{\partial C_i}{\partial x} \right)_{x_0},$$ (1)

where C_i is the impurity concentration in the liquid zone, U_0 is the zone velocity, D_i is the diffusion coefficient of the impurity in the liquid zone

$$\left(\frac{\partial C_i}{\partial x} \right)_{x_0} = \frac{C_i/x_0 - C_i}{\delta_i},$$

where $C_i/x_0 = kC_0 + (\partial C/\partial T)\Delta T$, k is the effective impurity-partition coefficient, ΔT is the supercooling, and δ_i is the breadth of the supercooling region.

At the liquid-solid interface we have, under equilibrium conditions, $U_0 = 0$ and

$$W = - \frac{\partial D_i}{\delta_i} (C_i/x_0 - C_0),$$ (2)

i.e., the width of the supercooling region is a function of D_i and the degree of intermixing in the melt. The width δ_i may be reduced by lowering the zone velocity U_0, but then the diffusion of impurity from the solid part of the bar into the liquid part cannot be neglected. After solidification of the liquid layer with the excess (or deficient) impurity content, the supercooling ΔT again

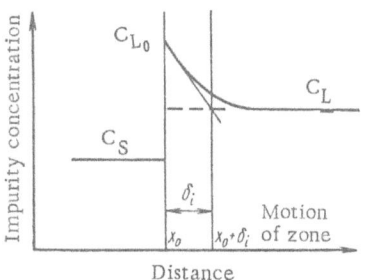

Fig. 1. Impurity distribution in the solid and liquid phases during zone recrystallization (k < 1).

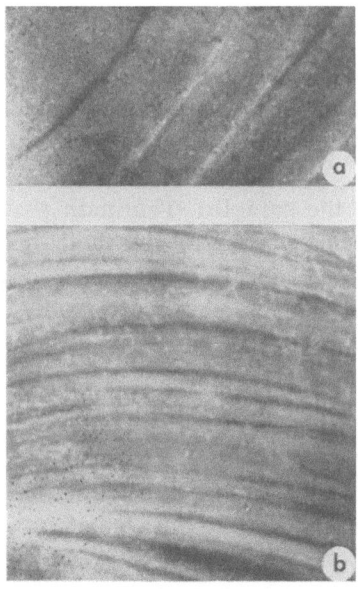

Fig. 2. Autoradiogram of the vertical (a) and horizontal (b) cross sections of a beryllium bar containing nickel isotope Ni[63] (× 10).

reaches a value sufficient to maintain a process of non-spontaneous crystallization, and the original growth velocity of the crystals is established. For each new spontaneous acceleration of dendrite growth, impurities are captured mechanically as a result of the high growth rate in the interdendrite space, and this brings the effective impurity-partition coefficient closer to unity, i.e., the efficiency of refining is reduced. Contact autoradiography was used to study the shape of the crystallization front during the electric-arc zone melting of niobium and beryllium.

The apparatus and method used for electric-arc zone melting are described in [1, 2]. We carried out such zone melting for niobium with a radioactive tungsten isotope, W^{182}, and for beryllium with nickel isotope Ni^{63}.

We see from the autoradiograms shown in Fig. 2 that the shape of the crystallization front during electric-arc zone melting is far from plane, i.e., from the condition on which the mathematical theory of zone refining is based. The position of the crystallization front in the liquid and solid parts of the bar is shown in Fig. 3. The variation in the crystallization angle between the direction of crystallization and the direction of zone motion creates conditions for a nonuniform distribution of impurities over the cross section of the bar. The complex parabolic form of the crystallization front (with respect to the vertical axial section) leads, first, to a change in the impurity capacity of individual layers of the zone when the melt is incompletely mixed (as in our case), and, secondly, to complication in the shape of the growing crystals. These conditions, together with the gravitational separation of impurities (the effect of which increases with falling zone velocity) and the impoverishment of the upper layers resulting from evaporation at the surface, leads to the enrichment of the lower layers with impurities for $k > 1$. The crystallization front may be described by the equation

$$f(r) = Az^p,$$

where A and p are constant and z is the coordinate of the layer with respect to the depth of the zone.

Then the ordinary equation for the distribution of impurities after one pass of the zone takes the form

$$C(x) = C_0\left[1 - (1 - k)\,e^{-\frac{kx}{l}}\right]. \tag{2}$$

In the case considered this may be rewritten in the form

$$C(x,\,z) = C_0\left[1 - (1 - k)\,e^{-\frac{k(x_0 + Az^p)}{l_0 - 2Az^p}}\right], \tag{3}$$

TABLE 1. Variation in the Content of Tungsten Isotope W^{182} in a Niobium Bar after Zone Melting (10 Zone Passes, k > 1)

Number of part	Distance from beginning of bar, mm	Radiation intensity, ×64 pulses/(min·cm²)	
		top of bar	bottom of bar
1	7	812	800
2	18	575	710
3	43	408	575
4	65	340	480
5	88	278	410
6	116	232	370
7	138	210	310
8	153	190	290
9	175	175	245

TABLE 2. Variation in the Concentration of Nickel Isotope Ni^{63} in a Beryllium Bar after Zone Melting (9 Zone Passes, k > 1)

Number of part	Distance from beginning of bar, mm	Radiation intensity, ×64 pulses/(min·cm²)	
		top of bar	bottom of bar
1	15	310	—
2	35	265	—
3	47	—	335
4	65	212	—
5	68	—	225
6	87	170	—
7	95	—	215
8	110	85	—
9	120	—	195
10	130	37	—
11	140	—	182
12	160	—	160

where l_0 is the length of the zone at the sample surface, and x_0 is the position of the origin of the zone in the bar at an arbitrary instant of time. Assuming an additive effect for successive zone passes (3), we have for n passes:

$$C_n(x,\ z) = C_0 \left[1 - (1-k)\, e^{-\frac{k(x_0 + Az^p)}{l_0 - Az^p}} \right]^n.$$ (4)

Expanding the exponential in Eq. (4) in series and neglecting terms containing k^2 (k < 1), we obtain

$$C_n(x,\ z) = C_0 k^n \left(1 - \frac{x_0 + Az^p}{l_0 - 2Az^p} \right).$$ (5)

The quantity $C_n(x, z)$ rises as $2Az^p \rightarrow l_0$, i.e., the depth of purification should increase on account of the change in zone length, but the efficiency should decrease. The latter factor predominates in the case considered, since k > 1. The effective impurity-partition coefficient k found from Eq. (5) equals

$$k = \sqrt[n]{\frac{C_n(x,z)}{C_0 \left(1 - \frac{x_0 + Az^p}{l_0 - 2Az^p} \right)}}.$$

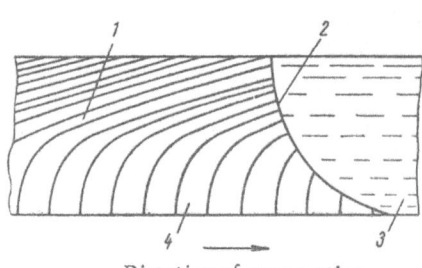

Direction of zone motion

Fig. 3. Position of crystallization
front and crystallizing grains during
the zone melting of beryllium: 1)
Growing grains; 2) crystallization
front; 3) liquid phase; 4) solid phase.

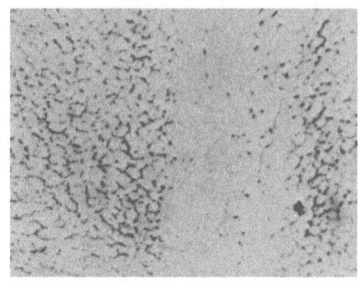

Fig. 4. Crystallization pulses in a
beryllium sample during zone melting (after 4 zone passes, bar length
160 mm, distance of section from
beginning of bar 110 mm, horizontal section of bar; ($\times$60).

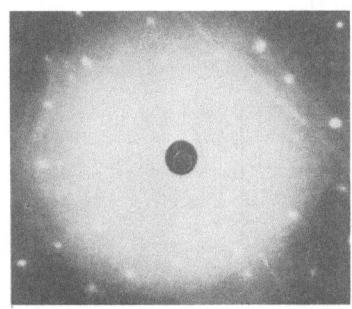

Fig. 5. Laue x-ray photograph of
a niobium single crystal obtained
by arc zone melting at a velocity of
0.75 mm/min, after 8 zone passes.

Thus the distribution of impurity along the bar
must be different in the upper and lower layers,
and so must the effective impurity partition coefficients for the upper and lower layers. This
is confirmed by data relating to the redistribution of tungsten in niobium (Table 1) and nickel
in beryllium (Table 2) in the upper and lower
layers of samples after electric-arc zone melting.

The partition coefficient for tungsten in
niobium varies from 6.25 in the upper layers to
4.0 in the lower.

Crystallization pulses ("jerky" crystallization) were also studied by the autoradiography
method. Metallographic study of these showed
that with increasing impurity concentration along the sample the crystallization pulses became
more frequent but less extended (Fig. 4). At 20 mm from the beginning of the beryllium bar
(bar length 180 mm), the average distance between pulses was 0.55 mm, and at the end (or rather 28 mm from the end) 0.18 mm.

All this supports what has been said regarding the role of structural supercooling. A
necessary condition for the formation of dendrites is the existence of a certain measure of
supercooling of the melt in advance of the crystallization front. The degree of supercooling
depends on the velocity of the crystallization front and on the thermal conductivity of the solid
phase. We noted that the zone velocity had an effect on the formation of single-crystal blocks
in the bar. At a velocity of 30 mm/min, dendrites with texture in the direction of zone motion
formed in the niobium bar. Despite the extremely unstable conditions at the crystallization
front, owing to the large number of crystallization centers in the lower and side regions of the
bar in contact with the water-cooled copper base, a zone velocity of 0.75 mm/min in niobium
was sufficient for the formation of single-crystal blocks (Fig. 5). In beryllium, the velocity of
0.75 mm/min was insufficient for the formation of single-crystal blocks in the bar, since the
thermal conditions at the crystallization front differed greatly from those corresponding to niobium, owing to the great thermal conductivity of beryllium.

Examination showed that in beryllium the (001) plane was almost perpendicular to the
grain-growth direction and only deviated slightly from a plane parallel to the crystallization

front. For noncrucible zone melting of beryllium, however, a velocity of 0.70 mm/min was sufficient for the formation of single-crystal sections.

Conclusions

1. The crystallization front in electric-arc zone melting is of complicated form.

2. As a result of the complex form of the crystallization front, the impurity distribution varies between the top and bottom of the bar in zone melting.

3. The zone velocity exerts an influence on the formation of single-crystal regions in the bar.

Literature Cited

1. Evstyukhin, A. I., et al. Collection of Scientific Papers on the Theory of Heat Resistance in the Academy of Sciences, USSR, Moscow, Metallurgizdat, 1961, p. 242.
2. Evstyukhin, A. I., et al. In collection: Metallurgy and Metallography of Pure Metals, No. 4, Moscow, Gosatomizdat, 1963, p. 69.
3. Vigdorovich, V. N., and Ivleva, V. S. Izv. Akad. Nauk SSSR, Otd. Tekhn. Nauk, Met. i Toplivo, No. 6:51 (1960).

DETERMINATION OF IMPURITIES IN HIGH-PURITY NIOBIUM

N. N. Kuznetsova and L. S. Krauz

Nature has many minerals containing niobium, but owing to the difficulty of separating this from its companion elements (tantalum, iron, manganese, zirconium, titanium, and thorium, uranium, tin, antimony, bismuth, tungsten, etc.) not many of these are of commercial significance. Among the principal minerals containing niobium are the tantaloniobates, constituting the salts of tantalic and niobic acids (for example, columbite, containing approximately 63% Nb_2O_5, 10% Ta_2O_5, 6% FeO, 15% MnO, 0.5% InO, 0.5% TiO_2), and the titanoiobates, salts of titanic and niobic acids, the most important among which are loparite (Na, Ca, Ce...)$_2 \cdot$ (Nb, Ti...)$_2O_6$ and pyrochlore (Na, Ca, Ce...)$_2 \cdot$ (Nb, Ti...)$_2O_6 \cdot$ [F, Pb].

The extraction of niobium from ores in the form of powder by carbon reduction or by means of such active metals as calcium, sodium, or magnesium belongs to the earlier period of metallic niobium production. More recently the electrolytic method of producing the metal has come into wider use, together with electron-beam melting. Quite recently (1957-1958) a technology has been developed in the Soviet Union and elsewhere for refining niobium by both arc and induction methods in vacuum or a neutral medium containing argon or helium. These methods yield a higher quality metal. According to chemical analysis, high-purity impurities (wt.%): $Zr - 2.5 \times 10^{-2}$, $Al - 5 \times 10^{-3}$, $Ti - 5 \times 10^{-3}$; trace of Cu, Fe, Mn, Mo, Ni, Mg, Ag; Si, Pb, Ca absent [1].

Niobium possesses such valuable properties as a high degree of resistance to various chemical actions (insolubility in mineral acids and alkalis) and high strength in its compounds. Until recently niobium was used mainly as an additive for raising the hardness of alloys, their corrosion resistance, nitriding rate, and so forth.

In recent years niobium has been more widely used in electrovacuum technology, semiconductors, and the production of cathode tubes (replacing tungsten and tantalum); its alloys have found extensive use in improving the quality of constructional material for the chemical industry and in reactor building.

The degree of purity of the metal governs the technological properties of niobium components, and the question of determining impurities is therefore of great importance. A number of papers have been devoted to this subject.

Direct Spectral Methods. The method proposed by A. A. Baskin and E. I. Zakharov [2] enables iron, silicon, titanium, and lead impurities to be determined with a sensitivity of $1 \times 10^{-3}\%$ (for tantalum $3 \times 10^{-2}\%$); the method described by G. M. Moroshkin and G. F. Malinin [3] gives only two impurities, aluminum and silicon in Nb_2O_5.

TABLE 1. Analytical Lines and Sensitivity of the Spectral
Determination of Impurity Elements in Concentrates Based
on Strontium Sulfate in the Analysis of Niobium

Element	Analytical line, A	Sensitivity of method ($k_{tot}=5$),%	Element	Analytical line, A	Sensitivity of method ($k_{tot}=5$),%
Mg*	2795.5	$5 \cdot 10^{-4}$	Co	3412.3	$2 \cdot 10^{-3}$
Ca*	4226.7	$2 \cdot 10^{-3}$	Ni	3050.8	$6 \cdot 10^{-4}$
Ba	2335.3	$4 \cdot 10^{-3}$	Cu*	3273.9	$4 \cdot 10^{-5}$
Al*	3082.2	$2 \cdot 10^{-3}$	Cd	2288.0	$6 \cdot 10^{-5}$
Ti	3088.0	$2 \cdot 10^{-3}$	Pb*	2831.1	$3 \cdot 10^{-4}$
Cr	2835.6	$1 \cdot 10^{-3}$	Bi	3067.7	$3 \cdot 10^{-4}$
Mn	2794.8	$2 \cdot 10^{-5}$	Sb	2598.1	$6 \cdot 10^{-4}$
Fe*	2720.6	$6 \cdot 10^{-4}$			

* The sensitivity of determining the element is indicated with due allowance for possible impurities in the reagents.

Chemical Methods. The determination of iron, manganese, molybdenum, tungsten, copper, nickel, zirconium, lead, phosphorus, tin, tantalum, and titanium impurities [4] is most frequently carried out from individual portions. In view of the fact that the presence of the niobium interferes with the determination of the majority of the elements, the authors of [5] first separated lead, cadmium, and bismuth by extracting them in the form of diethyldithiocarbamates with chloroform from alkaline solution; the author of [6] precipitated impurities in metallic niobium with thiacetamide, which has been quite widely mentioned in the literature as an analytical reagent for determining microscopic quantities of tin, lead, cadmium, and bismuth as a substitute for H_2S. The sensitivity of the method is 5×10^{-4}%.

Chemico-Spectral Methods. The chemico-spectral method developed by D. I. Ryabchikov et al. [7] for determining bismuth, gadolinium, antimony, tin, and lead impurities in metallic tungsten, niobium, and tantalum is based on the preliminary concentration of impurities by precipitating them in the form of sulfides on a CuS carrier and spectral determination of the concentrate with an ISP-22 spectrograph. The sensitivity of the method is 3×10^{-4} to 5×10^{-5}%.

A chemical-spectral method providing for the simultaneous determination of 15 impurity elements in metallic niobium is described below. The method is based on the chemical concentration of the elements with subsequent spectral analysis.

Thanks to the concentration of the impurities by the separation of the principal element (niobium), interference from the complex niobium spectrum is removed and the concentration of impurities in the concentrate is increased by a factor of five. The sensitivity of determining the elements is 2×10^{-3} to 6×10^{-5}% (Table 1), and the relative error of determination is 15 to 20%.

Choice of Method and Conditions for Concentrating Impurities

Analysis of metallic niobium is one of the most difficult of problems. In order to determine small quantities of impurities in high-purity materials two methods of concentration are commonly used: preliminary separation of the impurities with subsequent chemical or spectral determination, and concentration of micro-impurities after the removal of the principal element.

Concentration of impurities by means of group reagents (first variant) fails to produce the required result on account of the overlapping of the impurity and base (niobium) reactions.

The most suitable method is that of concentrating the impurities after separating out the principal element or base. Apart from the classical method (hydrolysis of niobium from

TABLE 2. Analysis of Artificial Mixtures Prepared from
720-mg Portions of Spectroscopically Pure Nb_2O_5,
Corresponding to 500 mg of Metallic Niobium,
by Introducing Impurity Elements

Elements	Content of elements, % of metallic-Nb portion					
	problem No. 1		problem No. 2		problem No. 3	
	introduced	found	introduced	found	introduced	found
Mg	$3.6 \cdot 10^{-3}$	$3.6 \cdot 10^{-3}$	$1.8 \cdot 10^{-3}$	$1.8 \cdot 10^{-3}$	$8.4 \cdot 10^{-4}$	$9.2 \cdot 10^{-4}$
Ca*	$2.2 \cdot 10^{-1}$	$2 \cdot 10^{-1}$	$2.2 \cdot 10^{-1}$	$2 \cdot 10^{-1}$	$2.2 \cdot 10^{-1}$	$2 \cdot 10^{-1}$
Ba	$1.3 \cdot 10^{-2}$	$1.2 \cdot 10^{-2}$	$1.2 \cdot 10^{-2}$	$1.1 \cdot 10^{-2}$	$8 \cdot 10^{-3}$	$6 \cdot 10^{-3}$
Al	$6 \cdot 10^{-3}$	$6.4 \cdot 10^{-3}$	$3 \cdot 10^{-3}$	$4 \cdot 10^{-3}$	$2.8 \cdot 10^{-3}$	$1.8 \cdot 10^{-3}$
Ti	$1 \cdot 10^{-2}$	$1.2 \cdot 10^{-2}$	$1.5 \cdot 10^{-2}$	$1.2 \cdot 10^{-2}$	—	—
Cr	$8.8 \cdot 10^{-3}$	$8 \cdot 10^{-3}$	$6 \cdot 10^{-3}$	$6.4 \cdot 10^{-3}$	$4 \cdot 10^{-4}$	$4 \cdot 10^{-4}$
Mn	$2.8 \cdot 10^{-4}$	$2.8 \cdot 10^{-4}$	$2.6 \cdot 10^{-4}$	$2 \cdot 10^{-4}$	$1.3 \cdot 10^{-4}$	$1.4 \cdot 10^{-4}$
Fe*	$1 \cdot 10^{-1}$	$7.5 \cdot 10^{-2}$	$1.3 \cdot 10^{-1}$	$1 \cdot 10^{-1}$	$1.3 \cdot 10^{-1}$	$1.2 \cdot 10^{-1}$
Co	$5.4 \cdot 10^{-3}$	$6.4 \cdot 10^{-3}$	$4.8 \cdot 10^{-3}$	$4.6 \cdot 10^{-3}$	$3.6 \cdot 10^{-4}$	$3.2 \cdot 10^{-4}$
Ni	$6 \cdot 10^{-3}$	$6 \cdot 10^{-3}$	$5 \cdot 10^{-3}$	$4.8 \cdot 10^{-3}$	$3 \cdot 10^{-3}$	$3.2 \cdot 10^{-3}$
Cu	$1.8 \cdot 10^{-3}$	$1.7 \cdot 10^{-3}$	$1 \cdot 10^{-3}$	$1 \cdot 10^{-3}$	$6 \cdot 10^{-4}$	$6.1 \cdot 10^{-4}$
Cd	$5.6 \cdot 10^{-4}$	$5.6 \cdot 10^{-4}$	$3.6 \cdot 10^{-4}$	$3.4 \cdot 10^{-4}$	$1.6 \cdot 10^{-4}$	$1.6 \cdot 10^{-4}$
Pb	$4.4 \cdot 10^{-3}$	$4 \cdot 10^{-3}$	$3.2 \cdot 10^{-3}$	$3.2 \cdot 10^{-3}$	$2 \cdot 10^{-3}$	$1.6 \cdot 10^{-3}$
Bi	$3.2 \cdot 10^{-3}$	$3.6 \cdot 10^{-3}$	$6 \cdot 10^{-3}$	$5.4 \cdot 10^{-3}$	$6 \cdot 10^{-4}$	$8 \cdot 10^{-4}$
Sb	$6 \cdot 10^{-3}$	$6 \cdot 10^{-3}$	$3 \cdot 10^{-3}$	$3.2 \cdot 10^{-3}$	$8.2 \cdot 10^{-4}$	$8 \cdot 10^{-4}$

*No tests involving the introduction of small quantities were made, as the element was present in the specimens.

acid solution), a very limited number of selective reactions for separating out large quantities of niobium are known. For determining small quantities of niobium, the following precipitating agents are most frequently used: tannin, Cupferron, pyrogallol, and 8-oxyquinoline [8-10]. These give bulky precipitates and in most cases require repeated reprecipitation. In order to separate out large quantities of niobium in compact form, the most effective method is that of extracting soluble niobium complexes directly from the solutions with organic solvents.

In view of the fact that niobium belongs to the class of easily hydrolyzed elements, not forming true solutions, extraction is effected from a strongly acid medium in order to avoid undesirable phenomena of polymerization and hydrolysis. As regards the extraction of large quantities of niobium, a limited number of extraction agents are described in the literature, of which the most widely used are diisopropylketone and benzoylphenylhydroxylamine.

The organic reagents described have some failings. Extraction of niobium with diisopropylketone is effected from a mixture of HCl and HF or H_2SO_4 and HF, and the use of HF is undesirable in view of the limited choice of vessels withstanding it [11].

Niobium can also be extracted by means of N-benzoylphenylhydroxylamine (BFGA) from strongly acid solutions. However, tantalum, zirconium, vanadium, and titanium are extracted as well [12].

After preliminary trials, we chose tributylphosphate (TBF) as extraction agent for niobium. The properties of this have been fully studied, and in recent years it has been widely used for the extraction of uranium, polonium, zirconium, and hafnium, which form complexes of almost stoichiometric composition with it.

Compared with ketones, TBF is more stable and requires no preliminary saturation with acid. Thanks to the presence of the P → O bond (phosphoryl group), which gives the extraction agent a strongly basic character, TBF should ensure a high extraction efficiency. TBF has the following advantages: It is not volatile (boiling point 289°C), so that its use is quite safe, it is almost insoluble in water (0.4 g/l at 20°C), it is extremely stable in the chemical respect (hydrolysis by water is practically eliminated, it is stable with respect to the action of many oxidizing agents), and finally its viscosity is 3.4 cp at 26 °C.

In order to reduce the viscosity and density of the organic phase and thus facilitate separation of layers, TBF is usually diluted with an inert hydrocarbon. There are no data regarding the extraction of large quantities of niobium from sulfuric-acid solution with tributyl-phosphate in the literature.

In order to simplify analysis, a sulfuric-acid medium was chosen for the extraction, since dissolution of niobium or Nb_2O_5 takes place in concentrated H_2SO_4. This eliminated the necessity of introducing supplementary operations before extraction (introduction of complexing reagents, citrates, tartrates, etc.). The possibility of introducing contamination was also removed in this way.

Preparation of the Sample for Analysis

It is well known that metallic niobium dissolves easily in hydrofluoric acid on the addition of drops of nitric acid. In the method proposed here for the analysis of niobium, the samples were dissolved in sulfuric acid after previously oxidizing the metal to the pentoxide at a temperature no higher than 600 °C, it having been found experimentally that Nb_2O_5 produced by roasting at a temperature exceeding 600 °C dissolved poorly in sulfuric acid. By using tributyl-phosphate as reagent, full and selective separation of niobium from impurities was achieved.

Using this method, a single extraction with a 60% solution of TBF in benzene extracted about 70% of the total amount of niobium, the partition coefficient being $D \approx 2$. In order to extract approximately 99.7% of the original niobium, five successive extractions with an equal volume of 60% TBF solution in benzene were required; the remaining 0.3% of the original niobium did not interfere with subsequent analysis. The completeness of the extraction of the impurities into the concentrate was confirmed by tests with artificial mixtures. For this purpose, various quantities of standard solutions of the elements were introduced into weighed amounts of spectroscopically pure Nb_2O_5 and the solutions were subjected to full analysis ($k_{tot} = 5$).* As may be seen from Table 2, the 15 impurities listed are concentrated almost completely (90 to 100%).

Analytical Procedure

The portion of metallic niobium (500 mg), previously oxidized to the pentoxide by roasting at 600 °C in a quartz dish for 2 h, or an equivalent amount of Nb_2O_5 720 mg in weight, is dissolved by strong heating in 90 ml of concentrated H_2SO_4.

The resulting sulfuric-acid solution of niobium is evaporated to a volume of 25 ml. To the cooled solution are slowly added 25 ml of H_2SO_4 (1 : 10), drop by drop from a graduated quartz dropping funnel, with careful mixing so as avoid hydrolysis of the niobium. The acidity of the solution after dilution corresponds to 20 N H_2SO_4. The cooled solution (50 ml of it) is transferred to a dry separating funnel, 50 ml of a 60% benzene solution of TBF are added and shaken up well for 7 min. In order to separate the phases, the solution is allowed to stand for

---

*In order to obtain spectroscopically pure Nb_2O_5, the capacity of a sulfuric solution of niobium to hydrolyze is used. To remove impurities, the Nb_2O_5 is dissolved in concentrated H_2SO_4. The resultant sulfuric solution of niobium (pH $\approx$ 2) undergoes hydrolysis, the niobium passing into the residue as niobic acid and the impurities mostly remaining in the filtrate. To obtain Nb_2O_5 of greater purity, the hydrolysis operation is repeated three to four times.

8 min. After separation of the layers, the aqueous phase is again transferred to a separating funnel and the extraction is repeated. The TBF extraction is carried out at a temperature not exceeding 20°C. After five extractions, no more than 0.3% of the total amount of niobium remains in solution. The last three extractions are therefore carried out with smaller portions of extraction agent: 30 ml of a 60% solution of TBF. Since partial extraction of the H_2SO_4 into the organic phase takes place, the acid concentration is made up after the third extraction by adding 2 ml of concentrated H_2SO_4 to the aqueous solution. The aqueous solution of impurities is evaporated until thick clouds of H_2SO_4 appear.

In order to decompose organic substances which pass partly into the aqueous phase during extraction, the solution is treated several times with 2 ml of concentrated HNO_3, with evaporation. When the solution has become completely colorless, the nitrates are decomposed and nitrogen-containing acids removed by treating the solution not less than three times with small quantities of water. The solution of impurities, concentrated to 2 or 3 ml, is transferred to a small quartz beaker 25 to 30 ml in capacity, to which have been previously added 210 ml $SrSO_4$, which corresponds to 100 mg of metallic strontium dissolved in 5 ml of concentrated H_2SO_4. The solution is evaporated to dryness and the dry residue roasted at 550°C for 1 h. The roasted residue is crushed and subjected to spectroscopic analysis.

A dummy run is made in parallel with the preparation of the concentrate in order to check the purity of all the reagents. A series of standards (five in all) is prepared by introducing standard solutions of all the elements into portions of spectroscopically pure strontium sulfate, which serves as a base for the standards and concentrates. All operations in the preparation of samples and standards are carried out in a quartz vessel. The acids and water are used only after previous distillation in a quartz apparatus. The tributylphosphate is purified by treating with soda and alkali and distilled in vacuum. The benzene is purified by distillation. The spectroscopically pure strontium sulfate is prepared from strontium nitrate.

Spectroscopic Determination of Impurities*

The spectroscopic analysis of niobium on a base of strontium sulfate was carried out by recording the spectra obtained from the chemical impurity concentrate and the standards, using a dc arc burning between carbon electrodes under identical conditions, on the same photographic plate. The electrodes were made of carbon rods 5 mm in diameter, the anode had a crater 4 mm deep and 4 mm in diameter, and the cathode was made in conical form with a 2-mm-diameter flat on the end. The electrodes were preliminarily annealed for 15 min and the spectra were excited in a dc arc at a current of 12 A and voltage 250 V. The photographs of the spectra were taken on a medium-dispersion quartz spectrograph (ISP-22, ISP-28) with a slit width of 10 μ, condenser diaphragm 1.2 mm, and exposure 2.5 min. A two-stage reducer was placed in front of the spectrograph slit, giving transmission steps of 100 and 10%.

After obtaining S_{L+B} (photographic density of line on background) and S_B (background) and having the characteristic curve of the photographic plate, we can find the logarithm of the intensity, introducing a correction for background at the same time. The calibrating curves are plotted in coordinates of (log I, log c). In calculating the results, the impurity-enrichment factor in the concentrate must be taken into account. (The analysis is effected quite simply without using a high-dispersion spectrograph. The chemico-spectral method here developed is being used for the analytical control of pure metallic niobium.)

Literature Cited

1.   Niobium and Tantalum [Russian translation], O. P. Kolchin (ed.), Moscow, IL, 1960, p. 307.

---

* The spectroscopic determinations were made by I. I. Smirenkina.

2. Baskin, A. A., et al. Zh. Anal. Khim., 16:627 (1961).
3. Moroshkina, G. M., and Malinin, G. F. Zh. Anal. Khim., 16:245 (1961).
4. Methods of Determining and Analyzing Rare Metals, Moscow, Izd. Akad. Nauk SSSR, 1961, p. 487.
5. Nazarenko, V. A., and Biryuk, E. L. Zavodsk. Lab., 25:28 (1959).
6. Yakovlev, P. Ya., et al. Zh. Anal. Khim., 17:90 (1962).
7. Ryabchikoy, D. I., et al. Reports of Analytical–Chemistry Commission, Moscow, Izd. Akad. Nauk SSSR, 12:82 (1960).
8. Alimarin, I. P., and Frid, B. I. Zavodsk. Lab., 7:109 (1938).
9. Savostin, A. P., and Alimarin, I. P. Vestn. Mosk. Gos. Univ., Khim., 1:45 (1960).
10. Hillebrand, V. F., et al. Practical Handbook on Inorganic Synthesis [Russian translation], 1957, p. 617.
11. Stevenson, P. C., and Hicks, H. G. Anal. Chem., 25:1517 (1953).
12. Petrukhin, O. M., and Chmutova, M. K. Abstracts of Contributions to the Conference on Extraction in Analytical Chemistry, Moscow, Izd. Akad. Nauk SSSR, 1961, p. 61.

INTERACTION OF SODIUM CHLORIDE WITH DYSPROSIUM AND GADOLINIUM CHLORIDES

K. V. Orlov, V. G. Kozlov, and N. G. Pospelova

The interaction of sodium chloride with dysprosium and gadolinium chlorides in melts is of great practical interest in the technology of producing metallic gadolinium and dysprosium by electrolysis from molten media, using a liquid cathode (zinc, cadmium), and also in producing gadolinium and dysprosium alloys with nonferrous and rare metals.

The systems formed by sodium chloride with dysprosium and gadolinium chlorides have never yet been studied.

The anhydrous chlorides of the rare-earth metals required for the investigation were obtained by dissolving gadolinium and dysprosium oxides in hydrochloric acid with subsequent boiling down of the hydrochloric solutions in vacuum in the presence of ammonium chloride.

For preparing the chlorides 99.85% gadolinium oxide and 99.7% dysprosium oxide were used.

The anhydrous chlorides obtained dissolved without turbidity in alcohol and water; they corresponded to the compositions 54.3% Gd + 45.03% Cl, 58.25% Dy + 41.37% Cl.

Chemically pure sodium chloride was first ground and melted. The interaction between the chlorides was studied by the fusion method, time-temperature curves being recorded with a Kurnakov pyrometer.

In view of the fact that gadolinium and dysprosium chlorides are easily hydrolyzed in the presence of moisture and atmospheric oxygen, the salts were melted and cooled in an argon-filled bomb.

The portion of salt mixture used for the thermograms was 2 to 3 g in weight, the cooling (or heating) rate was 5 to 10 deg/min, depending on the temperature range studied.

The NaCl − GdCl$_3$ System. The thermal-analysis results for this system are presented in Fig. 1 and Table 1.

The thermal analysis suggests that the system forms a chemical compound corresponding to the formula Na$_3$GdCl$_6$ (GdCl$_3$ · 3NaCl), which melts incongruently at 500°C and contains 40 wt.% NaCl. The chemical compound Na$_3$GdCl$_6$ and gadolinium chloride form a eutectic with a melting point of 398 ± 2°C at a NaCl content of 15 wt.%.

The NaCl − DyCl$_3$ System. The thermal-analysis results for this system are presented in Fig. 2 and Table 2.

K. V. ORLOV, V. G. KOZLOV, AND N. G. POSPELOVA

TABLE 1. Thermal-Analysis Data for the NaCl—GdCl₃ System

Salt content in mixture				Thermal effects, °C		
GdCl₃	NaCl	GdCl₃	NaCl	1	2	3
wt. %		mol. %				
100.0	—	100.0	—	670	—	—
95.0	5.0	98.8	1.2	570	—	395
90.0	10.0	97.5	2.5	470	—	395
85.0	15.0	96.2	3.8	—	—	395
80.0	20.0	94.75	5.25	—	470	396
70.0	30.0	91.45	8.55	—	498	398
65.0	35.0	89.4	10.6	—	495	388
60.0	40.0	87.1	12.9	—	500	395
50.0	50.0	81.9	18.1	605	500	—
40.0	60.0	75.0	25.0	680	498	—
30.0	70.0	65.8	34.2	720	500	—
20.0	80.0	53.2	46.8	750	495	—
10.0	90.0	32.9	67.1	795	—	—
—	100.0	—	100	800	—	—

TABLE 2. Thermal-Analysis Data for the NaCl—DyCl₃ System

Salt content in mixture				Thermal effects, °C			
DyCl₃	NaCl	DyCl₃	NaCl	1	2	3	4
wt. %		mol. %					
100.0	—	100.0	—	640	—	—	—
98.0	2.0	91.5	8.5	570	425	—	—
97.5	2.5	89.5	10.5	550	430	—	—
96.0	4.0	84.0	16.0	535	430	—	—
95.0	5.0	80.5	19.5	—	430	370	—
93.0	7.0	74.25	25.75	—	430	370	—
92.0	8.0	72.0	28.0	—	420	368	—
91.0	9.0	69.75	30.25	—	—	370	—
90.0	10.0	59.0	41.0	—	—	370	500
85.0	15.0	54.5	45.5	525	—	—	500
83.0	17.0	51.5	48.5	580	—	—	500
80.0	20.0	46.2	53.8	640	—	—	495
70.0	30.0	33.6	66.4	680	—	—	500
60.0	40.0	24.6	75.4	700	—	—	498
50.0	50.0	17.8	82.2	740	—	—	500
40.0	60.0	12.7	87.3	745	—	—	505
30.0	70.0	8.5	91.5	750	—	—	499
20.0	80.0	5.2	94.8	760	—	—	—
10.0	90.0	2.4	97.6	770	—	—	—
—	100.0	—	100.0	800	—	—	800

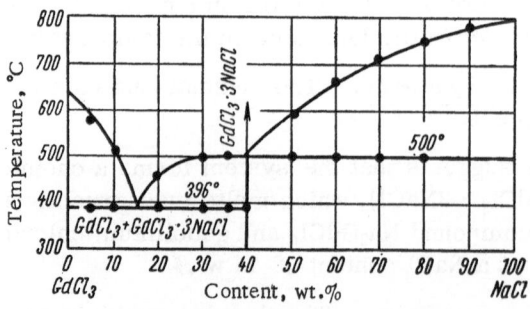

Fig. 1. Fusion diagram of the NaCl—GdCl₃ System.

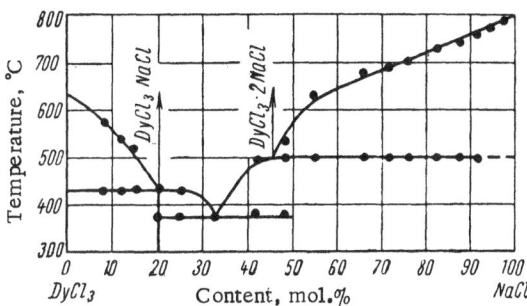

Fig. 2. Fusion diagram of the NaCl–DyCl₃ System.

The thermal analysis suggests that this system forms two chemical compounds. The first corresponds to the formula $NaDyCl_4$ ($DyCl_3 \cdot NaCl$) with a melting point of 430 ± 5°C and contains 19.5 mol.% NaCl; the second compound corresponds to the formula Na_2DyCl_5 ($DyCl_3 \cdot 2NaCl$), melts at 500 ± 2°C, and contains 45.5 mol.% NaCl. These chemical compounds form a eutectic with each other; the eutectic contains 30.25 mol.% NaCl and melts at 370 ± 2°C. In systems in which the interaction between the components leads to the formation of chemical compounds, a characteristic of the composition of the melt may be obtained theoretically from Schröder's equation. Despite the approximate nature of the calculation, some authors have used this method [1–3].

Our own calculation of the liquidus line for the NaCl–GdCl₃ system indicates agreement between the theoretical and experimental data in the presence of the complex $[GdCl_6]^{3-}$ ion in the system. The form of the liquidus curve for the NaCl–DyCl₃ system suggests the existence of $[DyCl_4]^{1-}$ and $[DyCl_5]^{2-}$.

Conclusions

1. We have studied the interaction of the components in the NaCl–GdCl₃ and NaCl–DyCl₃ systems by thermal analysis.

2. In the NaCl–GdCl₃ system an incongruently melting compound Na_3GdCl_6 with a melting point of 500°C is formed.

3. In the NaCl–DyCl₃ system there are two chemical compounds, corresponding to the formula $NaDyCl_4$ with a melting point of 430 ± 5°C and Na_2DyCl_5 with a melting point of 500 ± 2°C.

4. Calculations for the liquidus line (using the Schröder equation) indicates that complex ions of composition $[GdCl_6]^{3-}$, $[DyCl_4]^-$, and $[DyCl_5]^{2-}$ are probably present in the melts.

Literature Cited

1. Voskresenskaya, N. K. Izv. Sektora Fiz.-Khim. Analiza, Akad. Nauk SSSR, 23:155 (1953).
2. Markov, B. F., and Chernov, R. V. Ukr. Khim. Zh., 27:34 (1961).
3. Berg, L. G. Introduction to Thermography, Moscow, Izd. Akad. Nauk SSSR, 1961.
4. Ehrlich, P., et al. Z. Anorg. Allgem. Chem., 299:213 (1959).

EFFECT OF SLIGHT SILVER IMPURITIES ON THE ELECTRICAL RESISTANCE OF PALLADIUM

E. G. Okonnikov

It was pointed out earlier [1–3] that the electrical resistance and Hall constant of dilute solutions of silver in palladium reach peak values at a concentration of 0.13 at.% silver; so does the constant α defined by the equation $\Delta R_H/R = \alpha H^2$, where $\Delta R_H/R$ is the relative increment of resistance in a magnetic field H. Simultaneous solution of the galvanomagnetic and electric equations in the theory of Wilson and Sondheimer [4–6] enables us to calculate the effective number of conduction electrons (n_s) and holes (n_d) per cm^3, together with their mobilities τ_s/m_s and τ_d/m_d, where τ_i is the relaxation time and m_i is the effective mass for electrons (i = s) and holes (i = d). These calculations were carried out in [3].

By using these results and making certain assumptions, we can calculate the individual components of electrical resistance for a dilute solid solution of silver in palladium.

It is well known that the electrical resistance is zero in an ideal crystal lattice. Hence the actual resistance arises from disruptions to the regularity of the lattice. Main sources of such disruption include lattice vibrations, impurities, dislocations, etc. Hence the experimentally measured electrical resistance for nonmagnetic transition metals R_{exp} may be put in the form of a sum

$$R_{exp} = R_1(T) + R_2(T,\ s \to d) + R(c) + R(n_g) + \\ + R(n_3) + R_s + R_{el} + R(\Delta n^*) + \Delta R. \qquad (1)$$

In formula (1), $R_1(T)$ is the phonon part of the electrical resistance of the Grüneisen type [7] (this quantity may also be expressed in a more exact form, allowing for thermal expansion and the variation of the density of levels with temperature [8]);

$R_2(T, s \to d)$ is the part of the resistance only characterizing transition metals and depending on $s_1p \to d$ transitions. Subsequently, we shall only consider $R(T) = R_1(T) + R_2(T, s \to d)$. At room temperatures $R_1(T)$ and $R_2(T, s \to d)$ depend linearly on temperature, i.e.,

$$R(T) = R_1(T) + R_2(T,\ s \to d) = A_1 T + A_2 T = AT.$$

R(c) is the resistance associated with impurity atoms. According to Nordheim [12], $R(c) = B \cdot N^{1/3} c (1-c)$, where B is a constant made up of atomic constants, N is the concentration of impurity atoms per cm^3, and the c is the impurity concentration;

$R(n_g)$ is the resistance depending on the dislocation concentration. For an annealed sample $R(n_g)$ is a very small quantity [see estimate of $R(n_g)$ in [2]] at room temperature;

$R(n_3)$ is the resistance associated with scattering at grain boundaries. For fairly large grains (0.1 to 1 mm), $R(n_3)$ changes R_{exp} very slightly at fairly high temperatures [14];

R_S is the resistance associated with the diffuse scattering of electrons from the sample surface, R_{el} is the resistance due to interaction of the electrons; R_S and R_{el} play a considerable part at very low temperatures;

ΔR is a resistance including all other effects principally affecting the total resistance.

Since the resistance depends on the concentrations of electrons n_s and holes n_d, we shall in future consider that in formula (1) the electron and hole concentrations are in all cases the same as in the pure metal, and we shall allow for the corresponding variation in the electron and hole concentrations by a term $R(\Delta n^*)$.

The original palladium had a purity of 99.95%, and the samples contained the following amounts of silver impurity (at.%): No. 1 − 0.02; No. 2 − 0.07; No. 3 − 0.13; No. 4 − 0.23; No. 5 − 0.43; No. 6 − 0.63; No. 7 − 0.63; No. 8 − 1.02; No. 9 − 1.51 (see [1, 2] for more details on the alloys).

Since we are considering the region of room temperatures, the resistances $R(n_g)$, $R(n_3)$, R_S, R_{el}, and ΔR may be left out of the calculation. We may regard these as a slight constant increment to the phonon part of the electrical resistance, i.e.,

$$R^*(T) = R(T) + \sum R_i,$$

where $\sum R_i = R(n_g) + R(n_3) + R_S + R_{el} + \Delta R$. Then instead of Eq. (1) we obtain

$$R_{exp} = R^*(T) + R(c) + R(\Delta n^*). \tag{2}$$

In order to calculate each of the terms in formula (2), we may make use of the data of [3] and also make one or two assumptions.

Let us denote the mobilities of the electrons and holes determined by the vibrations of the crystal-lattice ions as $x_{0i} = \tau_{0i}/m_{0i}$ (i = s or d). Impurities will reduce their mobility by an amount $\Delta x_i = \tau_i^*/m_i^*$, this reduction corresponding to an increase in the electrical resistance in accordance with the law

$$R(c) = BN^{1/3}c(1-c) = B_0 c(1-c) = B_0 c^*.$$

[For dilute solid solutions $c^* = c(1-c) \simeq c$, i.e., there is a linear relationship.] The constant B_0 is easy to determine from the slope of the line relating R_{exp} to the concentration c of silver impurity (in the experimental relationships for the electrical resistance of a binary palladium-silver alloy [1, 2] with $c \cdot 100 \geq 0.43$ at.% silver, R_{exp} varies linearly with silver concentration). Knowing B_0, we can easily find the fall in the electrical resistance for a given impurity concentration as compared with the pure metal. Then

$$\frac{1}{R(c)} = \sigma(c) = e^2\left(n_s^* \frac{\tau_s^*}{m_s^*} + n_d^* \frac{\tau_d^*}{m_d^*}\right) = e^2 n^*\left(\frac{\tau_s^*}{m_s^*} + \frac{\tau_d^*}{m_d^*}\right). \tag{3}$$

Here $n^* = n_s^* = n_d^*$, since palladium is one of those metals with equal number of conduction electrons and holes [6, 13-15]. Knowing n^*, we can find the total mobility of electrons and holes $\Delta x = (\tau_s^*/m_s^*) + (\tau_d^*/m_d^*)$ determined by the impurity concentrations alone. Knowing Δx and x [3], we can easily calculate the mobility of the electrons and holes determined by the lattice vibrations alone:

$$x_0 = x + \Delta x = \left(\frac{\tau_s}{m_s} + \frac{\tau_d}{m_d}\right) + \left(\frac{\tau_s^*}{m_s^*} + \frac{\tau_d^*}{m_d^*}\right) = \frac{\tau_{0s}}{m_{0s}} + \frac{\tau_{0d}}{m_{0d}}.$$

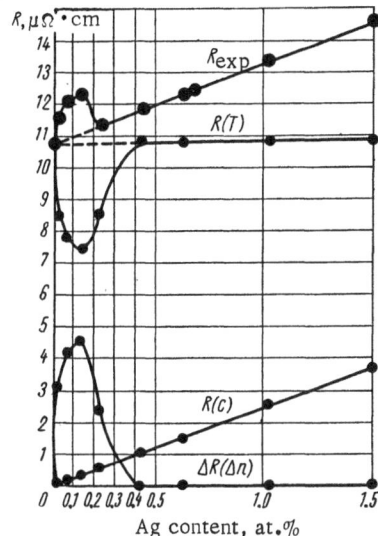

Resistances R_{exp}, $R(T)$, $R(c)$, and $\Delta R(\Delta n)$ as functions of impurity concentration.

If n_0^* is the effective number of conduction electrons for pure palladium ($n_0^* = n_{0S}^* = n_{0d}^*$), then $\sigma(T) = e^2 n_0^* x$ determines simply the electrical conductivity due to the lattice vibrations. Then $R^*(T) = 1/\sigma(T)$ is the phonon part of the electrical conductivity.

From the known R_{exp}, $R^*(T)$, and $R(c)$, we determine $R(\Delta n^*) = R_{exp} - R(T) - R(c)$. These resistances are given as functions of impurity concentrations in the figure. We see that $R(T)$ has quite a deep minimum at 0.13 at.% silver, while $R(\Delta n^*)$ has a marked maximum at the same concentration. These extrema may be explained by I. B. Borovskii and K. P. Gurov's theory of atomic hardening blocks.

According to this theory, if the crystal lattice contains an impurity ion with a charge in excess of that on the ions of the principal metal, then this ion becomes surrounded by a region* in which the ordinary metallic bond is supplemented by an additional bond of a polar type. The intensification of the bond in the metal lattice leads to a reduction in the vibration amplitude of the lattice ions and hence to a reduction in the phonon part of electrical resistance $R(T)$. The more the metal becomes filled with such "atomic blocks," the smaller will $R(T)$ be. The metal is completely filled with blocks when these assume close packing. When the "atomic blocks" overlap, the additional bond of the polar type vanishes, and the amplitude of the lattice-ion vibrations rises to its earlier value when the blocks overlap completely; the resistance correspondingly resumes its value for the practically pure metal.

As a result of the distortion of the s and d energy levels around the impurity ion with the excess charge, there may be a transfer of electrons from the s to the d level, leading to a rise in the effective electron mass. The effective mass of the s electrons may also change.

The change in the effective masses of electrons and holes is equivalent to a reduction in their effective number. For finding the effective masses of the electrons and holes the three equations (Hall effect, resistance in a magnetic field, and resistance with no field) are insufficient; hence we shall only refer to the effective number of conduction electrons. Naturally, on reducing the effective number of electrons and holes the resistance $R(\Delta n^*)$ should rise.

As shown in [3], the effective number of conduction electrons (and, correspondingly, holes) actually does fall on increasing the silver content from 0 to 0.13 at.%, rising again on increasing the content further. For complete overlapping of the blocks (for c approximately equal to 0.43 at.% silver), $R(\Delta n^*)$ is close to zero, and $R(T)$ approximately equals the resistance of the pure metal.

It is possible (in some cases) that on transferring electrons from the d to the s level the effective number of electrons may rise (effective mass fall) and $R(\Delta n^*)$ will have a minimum value for close packing of the blocks (this corresponds to an increase in conductivity); then the $R_{exp} = f(c)$ curve will have a minimum instead of the maximum obtained for palladium-silver alloys [16-24].

---

* For silver and palladium ions this region has a diameter of about 28 A at 20°C [2].

Literature Cited

1. Okonnikov, E. G. Tr. Inst. Met., Akad. Nauk SSSR, No. 6:32 (1960).
2. Okonnikov, E. G. In collection: Metallurgy and Metallography of Pure Metals, No. 3, Moscow, Gosatomizdat, 1961, p. 295.
3. Okonnikov, E. G. In collection: Metallurgy and Metallography of Pure Metals, No. 4, Moscow, Gosatomizdat, 1963, p. 188.
4. Sondheimer, E. H., and Wilson, A. H. Proc. Roy. Soc., A190:435 (1947).
5. Sondheimer, E. H. Proc. Roy. Soc., A193:482 (1948).
6. Wilson, A. H. The Theory of Metals, Cambridge, The University Press, 1953.
7. Grüneisen, E. Ann. Phys., 16:530 (1933).
8. Birss, R. R., and Dey, S. K. Proc. Roy. Soc., A263:473, 1951.
9. Mott, N. F. Proc. Phys. Soc., 47:571 (1935); A153:699 (1936); A156:368 (1936).
10. Wilson, A. H. Proc. Phys. Soc., A167:580 (1938).
11. Seitz, F. Modern Theory of Solids [Russian translation], Moscow, IL, 1949. [English edition: New York, McGraw-Hill, 1940.]
12. Nordheim, L. W. Ann. Phys., 9:607 (1931).
13. Borovik, E. S. Zh. Éksperim. i Teor. Fiz., 23:83 (1952); 27:362 (1954); 30:262 (1956).
14. Borovik, E. S., and Volotskaya, V. G. Zh. Éksperim. i Teor. Fiz., 36:1650 (1959).
15. Borovik, E. S., and Volotskaya, V. G. Fiz. Metal. i Metalloved., 6(1):60 (1958).
16. Lifshits, V. G. Physical Properties of Metals and Alloys, Moscow, Metallurgizdat, 1959.
17. Borovskii, I. B., and Gurov, K. P. Fiz. Metal. i Metalloved., 4:10 (1957).
18. Borovskii, I. B., and Gurov, K. P. Study of Heat-Resistant Alloys, Vol. 2, Moscow, Izd. Akad. Nauk SSSR, 1957, p. 251.
19. Borovskii, I. B., and Gurov, K. P. Fiz. Metal. i Metalloved., 7:225 (1959).
20. Borovskii, I. B., and Gurov, K. P. Zh. Éksperim. i Teor. Fiz., 36:1203 (1959).
21. Borovskii, I. B., and Gurov, K. P. Dokl. Akad. Nauk SSSR, Ser. Fiz., 23:660 (1959).
22. Borovskii, I. B., and Gurov, K. P. Fiz. Metal. i Metalloved., 10:513 (1960).
23. Gurov, K. P. Nauchn. Dokl. Vysshei Shkoly, Fiz.-Mat. Nauki, 1:152, 159 (1958).
24. Gurov, K. P. Tr. Inst. Met., Akad. Nauk SSSR, No. 6:19 (1960).

SOME LAWS REGARDING CHANGES IN THE PARAMETERS D_0 AND Q DURING DIFFUSION IN METALS AND ALLOYS

G. B. Fedorov

In 1952 the author obtained an experimental relationship between the pre-exponential factor D_0 in the diffusion equation and the activation energy Q. This relationship may be expressed in the form

$$D_0 = A \cdot e^{Q/B},\tag{1}$$

where A and B are constants. A relation of this kind was first derived for the diffusion of chromium in nickel-base solid solutions on the assumption of approximate equality between the diffusion coefficients at a temperature close to the melting points of the alloys [1]. It was shown that the relation was valid for diffusion in alloys of the iron groups also. At the same time it was noted that the equality of the coefficients A and B in the analysis of different cases of diffusion indicated that the same diffusion mechanism held in all such cases.

Analysis of the results of approximately 40 cases of diffusion in solid solutions of the substitution type in alloys of iron and cobalt with fcc lattices showed that the relation between D_0 and Q could be described by means of the same coefficients A and B [2, 3]. If the data of [2] are supplemented by information on the diffusion of chromium and nickel in nickel-chromium solid solutions [4] and all the values of D_0 and Q are plotted on a graph $\log D_0 = \varphi(Q)$, then to a fair degree of experimental accuracy the relationship between D_0 and Q may be expressed by means of the same coefficients A and B for every one of the alloys (Fig. 1).

Subsequently it was found that an analogous relationship between D_0 and Q also existed for the diffusion of carbon in steels, i.e., in cases in which the diffusion mechanism was of an interstitial nature [5]. As we should expect, the values of A and B in this case differed from those characterizing the vacancy mechanism [6]. Table 1 presents a set of data on the diffusion of carbon in alloys of α iron [5], together with Wert's results on the diffusion of carbon in α iron [7] and other data on the diffusion of carbon in γ iron [5, 8–10] and nickel alloys [11]. If we plot a graph of $\log D_0$ against Q from these data (Fig. 2), we notice that the whole graph is divided into three groups of points. Each group has its own rectilinear relationship between $\log D_0$ and Q, the slopes of the three straight lines, i.e., the coefficients B (Table 2), being approximately equal. We also note that the straight lines for γ iron and nickel are very close to one another, so that the coefficients A also differ only slightly. This last fact is due to the similarity between the crystal lattices of γ iron and nickel. Part of the displacement between the lines is probably attributable to the different lattice parameters of the elements. A change in the type of crystal lattice leads to a considerable change in coefficient A (by something like two orders).

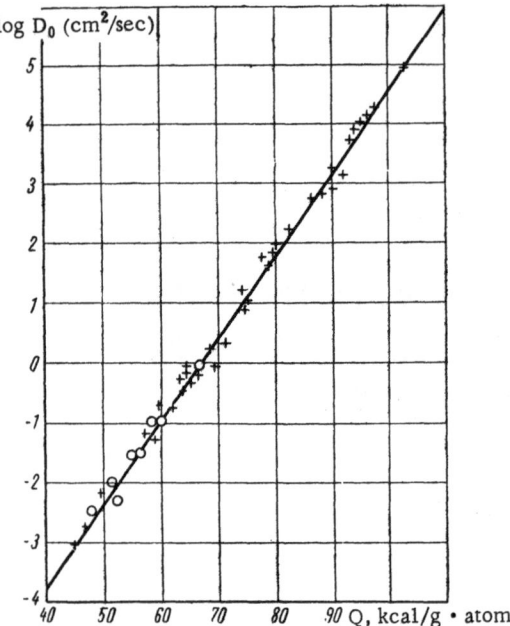

Fig. 1. Graph relating D_0 to Q for alloys based on iron-group metals for the diffusion of metals forming solid solutions of substitution with the former: +) Data from [2]; ○) data from [4].

The fall in the activation energy for the diffusion of carbon on passing from the fcc lattice of γ iron and nickel to the bcc lattice of α iron may be explained by the size of the interstitial vacancies in which the carbon atom lies. Both in fcc and bcc lattices, these vacancies are octahedral, though distorted (oblate) in the bcc lattices. We know that the radius of an octahedral vacancy in close packing is 0.41 of an atomic radius; the minimum radius of an octahedral vacancy in the bcc lattice, however, is only 0.15 of an atomic radius. Even in the fcc lattice, the carbon atom has to push the metal atoms away, slightly distorting the lattice. In the bcc lattice, however, the octahedral vacancy is much smaller and the distortion of the crystal lattice considerably greater, which produces a lower lattice and reduces the activation energy of diffusion. In addition to this, the size of the vacancy in the bcc lattice is so small that the carbon atom has to free itself from some of the valence electrons to a greater degree than in the fcc lattice. Hence the charge on the positive carbon atom is greater for diffusion in a bcc lattice than in an fcc lattice. This conclusion is in full agreement with information obtained on the electrodiffusion of carbon in α and γ iron and nickel in [12, 13].

A tentative explanation for the $D_0 : Q$ relationship here described may be based on the Wert-Zener-Le Claire diffusion theory [7, 14, 15]. According to this theory, the pre-exponential factor D_0 for diffusion in interstitial solid solutions may be put in the form

$$D_0 = \gamma a^2 \nu e^{\frac{\Delta S}{R}}, \qquad (2)$$

where ΔS is the activation entropy of diffusion, a is the lattice constant, ν is the vibration frequency of the diffusing atom at its equilibrium position, γ is a geometrical factor determined by the type of crystal lattice in the solvent, and R is the gas constant. For an fcc lattice, γ equals unity, and for a bcc lattice, $\frac{1}{6}$.

Supposing that the potential energy of the atom of the dissolved element varies sinusoidally during its motion from one interstitial position to another, with an amplitude equal to the activation energy Q, the frequency ν equals

$$\nu = \left(\frac{Q}{2m\lambda}\right)^{1/2}, \qquad (3)$$

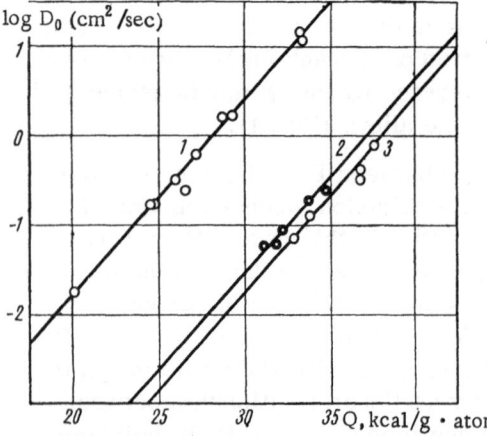

Fig. 2. Relation between D_0 and Q for the diffusion of carbon in iron- and nickel-base alloys: 1) Alloys of α iron; 2) alloys of γ iron; 3) alloys of nickel.

TABLE 1. Diffusion Parameters of Carbon in Iron
and Nickel Alloys

Base of alloy	Impurity content, wt. %	Q, kcal/g · atom	D_0, cm²/sec	Literature reference
α-Fe	—	20.1	0.02	[7]
	—	24.6	0.2	[5]
	0.79 Si	26	0.4	[5]
	2.38 Si	27.2	0.8	[5]
	3.6 Si	29.3	2.2	[5]
	0.46 Ni	24.8	0.2	[5]
	2.0 Ni	26.6	0.3	[5]
	0.56 Mo	28.9	2.0	[5]
	2.58 Mo	33.5	20.0	[5]
	0.93 Cr	33.6	16.4	[5]
γ-Fe	—	32.4	0.1	[5]
	—	32	0.07	[8]
	—	31.35	0.07	[9]
	0.3 C	35	0.3	[10]
	0.5 C	34	0.22	[10]
Ni	—	33	0.08	[11]
	0.74 Cr	34	0.15	[11]
	4.65 Cr	37	0.5	[11]
	5.25 Co	37	0.4	[11]
	2.94 Mo	38	1.0	[11]

TABLE 2. Values of Coefficients A and B for Diffusion
in Iron–Group Alloys

Material in which diffusion takes place	Diffusion element	Type of solid solution	A, cm²/sec	B, cal/g · atom
α-Fe	C	Interstitial	$6 \cdot 10^{-7}$	1950
γ-Fe	C	"	$1 \cdot 10^{-8}$	2000
Ni	C	"	$5 \cdot 10^{-9}$	1980
Fe, Co, Ni fcc	Fe, Co, Cr, Ni, W	Substitution	$6 \cdot 10^{-10}$	3090

where λ is the distance between the intermediate positions and m is the mass of the diffusing atom.

The activation entropy of diffusion may be determined from the expression

$$\Delta S = - Q \frac{\partial (\mu/\mu_0)}{\partial T} , \tag{4}$$

where μ is the shear modulus of the crystal lattice at temperature T, and μ_0 is the modulus at 0°K.

The latter relationship was derived on the following simplifying assumptions: a) It was assumed, without allowing for the effect of relaxation processes, that the shear modulus varied linearly with temperature, and b) the whole work associated with the motion of the diffusing atom from the equilibrium position to the top of the potential barrier separating it from the next equilibrium position could be reduced to the elastic energy concentrated in the lattice surrounding the diffusing atom when the latter lay at the top of the barrier. This work equals the change in the free energy ΔG, which is given by the relation $\Delta G = Q - T\Delta S$.

Putting expression (4) into (2), we obtain

$$D_0 = \gamma a^2 \nu e^{-\frac{Q}{R} \frac{\partial (\mu/\mu_0)}{\partial T}} . \tag{5}$$

Thus we obtain the same D_0 : Q relationship as that found experimentally. The coefficients A and B in (1) are determined in the following way from (5):

$$A = \gamma a^2 \nu, \quad B = - \frac{R}{\frac{\partial (\mu/\mu_0)}{\partial T}} . \tag{6}$$

Let us make some approximate calculations for the numerical values of A and B obtained from this theory. If we suppose that the variation of the shear modulus with temperature is analogous to that of the Young's modulus, we may use Köster's [16] that to determine $\partial (\mu/\mu_0)/\partial T$. For α iron this quantity is approximately equal to 3×10^{-4}. Hence the value of B equals 6600, which is almost three times that given in Table 2. This is a considerable difference when we remember that B comes in the index of the exponential.

Let us try some refinements to the calculation. First of all, we note that the assumption of a linear relationship between shear modulus and temperature is in general not justified. As a result of relaxation processes, the elastic moduli in fact fall faster with increasing temperature than they would by the law

$$\mu = \mu_0 + T (\partial \mu / \partial T), \tag{7}$$

where the derivative $(\partial \mu / \partial T)$ is regarded as constant. As temperature rises, the absolute value of the derivative of shear modulus with respect to temperature in fact becomes greater. We may provisionally suppose that the second derivative of the modulus with respect to temperature remains constant as temperature varies. Then the temperature dependence of the shear modulus may be written

$$\mu = \mu_0 + \left(\frac{\partial \mu}{\partial T} \right)_0 \cdot T + \frac{\partial^2 \mu}{\partial T^2} \cdot T^2, \tag{8}$$

where the zero indices refer to absolute zero temperature. From this we obtain

$$\Delta S = - Q \left\{ \left[\frac{\partial (\mu/\mu_0)}{\partial T} \right]_0 + \frac{\partial^2 (\mu/\mu_0)}{\partial T^2} T \right\} . \tag{9}$$

We notice that the activation entropy of diffusion contains a term including the temperature. It is interesting to note that temperature also comes into one of the terms in the expression for the activation entropy of the process underlying the development of centers of "local melting" in a solid metal (K. A. Osipov [17]). Osipov also uses this expression to describe the self-diffusion process.

On calculating the coefficient B for the diffusion of carbon in α iron (allowing for relaxation processes) from the same data [16] for the normal elastic moduli, we obtain a value of approximately 2×10^3, which practically coincides with the experimental value given in Table 2.

The numerical value of coefficient A calculated from formula (6) for the diffusion of carbon in α iron comes out at approximately 10^{-4}. In Table 2, however, this coefficient is nearly three orders smaller. This large difference may be explained on the basis of the following considerations.

The basis for the calculation of the diffusion coefficient is the expression

$$D = \alpha \frac{a^2}{t} , \tag{10}$$

where α is a geometrical factor depending on the arrangement of neighboring nodes to which an atom may pass around the particular node at which it lies in its equilibrium state, t is the average time between two successive leaps of each atom. The assumption that after each jump any atom moves successively away from its original position cannot be regarded as correct. The atoms may return, either at once or after a certain amount of meandering, to their original position [15, 18-20] or one close to it, so that we must replace t by a certain t_{eff}:

$$t_{eff} = \frac{t}{\beta}, \tag{11}$$

where β is the correlation coefficient. This coefficient is always smaller than unity, and may clearly also be called the transmission coefficient as in the theory of absolute reaction velocities [21]. Then expression (10) may be rewritten in the form

$$D = \alpha\beta \frac{a^2}{\tau}. \tag{12}$$

After appropriate transformations, the correlation coefficient turns up in the expression for A:

$$A = \beta\gamma a^2 \nu. \tag{13}$$

This correction brings the A value given by formula (13) into agreement with the values given in Table 2. The value of the correlation coefficient is of the order of 10^{-3} or smaller.

Similar considerations may be applied to the analysis of the diffusion of elements forming substitution-type solid solutions with the solvent. Figure 1 and Table 2 present data relating the diffusion in γ iron, cobalt, and nickel, which have fcc lattices and differ only slightly in lattice constant. It should be noted that data relating to the self-diffusion of these elements are also shown in Fig. 1 and fall quite neatly on a single straight line. Hence the A and B coefficients in this case imply a relation between the self-diffusion parameters. As indicated by Le Claire [15], the following relation is satisfied for self-diffusion:

$$\Delta S = -0.55 \cdot Q \frac{\partial(\mu/\mu_0)}{\partial T}. \tag{14}$$

If we consider the relaxation processes taking place in the metal, then in analogy with Eq. (9) the activation entropy for self-diffusion may be rewritten in the form

$$\Delta S = -0.55 \cdot Q\left\{\left[\frac{\partial(\mu/\mu_0)}{\partial T}\right]_0 + \frac{\partial^2(\mu/\mu_0)}{\partial T^2}T\right\}. \tag{15}$$

The coefficient B from expression (1) is given as

$$B = -\frac{R}{0.55\left\{\left[\dfrac{\partial(\mu/\mu_0)}{\partial T}\right]_0 + \dfrac{\partial^2(\mu/\mu_0)}{\partial T^2}T\right\}}. \tag{16}$$

The value of coefficient A is given by (13) in this case also, where $\gamma = 1$ for both fcc and bcc lattices.

Taking the numerical value of the expression in curly brackets in formula (16) as approximately the same for both types of solid solutions in iron, we can understand why the coefficient B is greater for substitution-type solid solutions than for the interstitial type. In the latter case the factor 0.55 is absent from the formula for B.

Thus relation (1) is valid both for self-diffusion and for the diffusion of elements forming substitution- and interstitial-type solid solutions with the solvent metal. We must of course stipulate that the constancy of A and B only applies to the case in which the quantities $[\partial(\mu/\mu_0)/\partial T]_0$ and $[\partial^2(\mu/\mu_0)/\partial T^2]T$ are not seriously altered in the course of alloying, in addition to $a^2\nu$. This is really only possible when the quantities of alloying element are comparatively small, or, more precisely, when alloying produces no serious change in the nature of the chemical bond.

It is interesting to note that expression (1) may also be obtained from V. Z. Bugakov's formula for the heterodiffusion coefficient [22]. This was mentioned by P. L. Gruzin in [2]. The coefficients A and B calculated from Bugakov's formula agree almost entirely with the experimental values of Table 2.

Equations (1) and (2) may be used to obtain a formula giving the activation energy of self-diffusion ΔS in terms of the coefficient B and the self-diffusion activation energy:

$$\Delta S = Q\,(R/B). \tag{17}$$

For nickel, formula (17) gives 42 cal/(g · atom · deg). This quantity is almost four times greater than calculated from formula (2) on the basis of the experimental pre-exponential factor D_0 for the self-diffusion of nickel.

It is appropriate to mention the fact that a similar high value of self-diffusion activation entropy may also be obtained from other considerations. In his book [17], Osipov proposed expressing certain activated processes in solid metals and alloys in terms of the activation energy q and entropy ΔS_a of the process underlying the development of a "local-melting" center in a solid metal. According to Osipov, the self-diffusion activation energy equals 3q for metals with an fcc lattice [23]. It would be entirely logical to express the self-diffusion activation entropy as $3\Delta S_a$. The author nevertheless refrained from doing this, since the values of the pre-exponential factor calculated from formula (2) are very large in this case. Finally the author did not take the correction in the form of the correlation coefficient into consideration. If, however, we calculate the self-diffusion activation entropy from the formula $\Delta S = 3 \cdot \Delta S_a$, we obtain 34 cal/(g · atom · deg) for nickel, which is quite close to the value given by expression (17).

Conclusions

1. The $D_0 : Q$ relationship of form

$$D_0 = A \cdot \exp{(Q/B)}$$

is universal. This expression is valid not only for the self-diffusion of metals but also for diffusion in interstitial and substitution-type solid solutions whenever the alloying process leaves the nature of the chemical bond unchanged.

2. Similar values of coefficients A and B indicate that similar diffusion mechanisms are involved. The value of A changes on changing the type of crystal lattice.

3. The relationship thus established may be explained on the basis of the Wert-Zener-Le Claire theory. In order to obtain quantitative agreement between this theory and experiment, certain corrections must be introduced for relaxation processes, and for the possible back diffusion of the atoms in the course of diffusion, which reduces the directional diffusion flow.

4. The validity of the laws found is also supported by the heterodiffusion theory of V. Z. Bugakov. The relatively large value of self-diffusion activation entropy may be explained on the basis of K. A. Osipov's self-diffusion theory after introducing appropriate corrections.

Literature Cited

1. Gruzin, P. L., and Fedorov, G. B. Dokl. Akad. Nauk SSSR, 105:264 (1955).
2. Gruzin, P. L. Probl. Metalloved. i Fiz. Metal., No. 4:475 (1955).
3. Kurdyumov, G. V. In book: Use of Isotopes in Technology, Biology, and Agriculture, Moscow, Izd. Akad. Nauk SSSR, 1955, p. 48. (Contributions of the Soviet delegation to the International Conference on the Peaceful Use of Atomic Energy, Geneva, 1955.)

4. Fedorov, G. V., et al. In collection: Metallurgy and Metallography of Pure Metals, No. 4, Moscow, Gosatomizdat, 1963, p. 110.

5. Gruzin, P. L., et al. Tr. Vses. Nauchn.-Tekhn. Konf. po Primeneniyu Radioaktivn. i Stabil'n. Izotopov i Izluchenii v Nar. Khoz. i Nauke, Moscow, 1957. Met. i Metalloved., 1958, p. 246.

6. Fedorov, G. V., and Gulyakin, V. D. In collection: Metallurgy and Metallography of Pure Metals, No. 1, Izd. MIFI, 1959, p. 170.

7. Wert, C. A. Phys. Rev., 79:601 (1950).

8. Wells, C., and Mehl, R. Trans. AIME, 140:294 (1940).

9. Blanter, M. E. Zh. Éksperim. i Teor. Fiz., 18:529 (1948).

10. Appleton, A. S. Trans. AIME, 230:893 (1964).

11. Gruzin, P. L., et al. Fiz. Metal. i Metalloved., 4:94 (1957).

12. Babikova, Yu. F., and Gruzin, P. L. Fiz. Metal. i Metalloved., 5:255 (1957).

13. Gruzin, P. L., et al. Radioisotopes in Scientific Research, Vol. 1, New York, Pergamon Press, Inc., 1958.

14. Wert, C. A., and Zener, C. Phys. Rev., 76:1169 (1949).

15. Le Claire, A. D. Usp. Fiz. Metal., No. 1:224 (1956).

16. Köster, W. Z. Metallk., 39:1 (1948).

17. Osipov, K. A. Some Activated Processes in Solid Metals and Alloys, Moscow, Izd. Akad. Nauk SSSR, 1962.

18. Le Claire, A. D. Phil. Mag., 3:921 (1958).

19. Manning, J. R. Phys. Rev., 116:819 (1959).

20. Manning, J. R. Phys. Rev., 124:470 (1961).

21. Glasstone, S., et al. Theory of Absolute Reaction Velocities [Russian translation], Moscow, IL, 1948.

22. Bugakov, V. Z. Diffusion in Metals and Alloys, Moscow, Gostekhizdat, 1949.

23. Osipov, K. A. Theory of the Heat Resistance of Metals and Alloys, Moscow, Izd. Akad. Nauk SSSR, 1960.

SELF-DIFFUSION IN ALPHA URANIUM

G. B. Fedorov, E. A. Smirnov, and F. I. Zhomov

In some papers [1–4] devoted to the study of self–diffusion in the α phase of uranium, main attention has been paid to questions of diffusion anisotropy and boundary diffusion. Thus a study of self–diffusion along the three principal crystallographic directions in [1] revealed no diffusion anisotropy, since the differences between the diffusion coefficients along the principal axes lay within the limits of experimental error. The order of the diffusion coefficients measured in this paper at 640°C was 10^{-14} cm²/sec. The authors of [2] considered that if any anisotropy existed it was negligible. The temperature dependence of the self–diffusion coefficient in polycrystalline uranium for the temperature range 580 to 650°C has the form $D = 2 \times 10^{-3} \exp(-40,000/RT)$ cm²/sec. It was shown radiographically that below 600°C the diffusion took place mainly along grain boundaries. A further study of self–diffusion along the three principal crystallographic directions at 625°C [3] showed considerable anisotropy of volume diffusion. The diffusion coefficients along the [100] and [001] directions were 1.95×10^{-13} cm²/sec, while for the [010] direction the coefficient of volume diffusion calculated from the initial part of the concentration curve was no greater than 10^{-14} cm²/sec. The authors of [3] explained the deviation of the $\log c_{[010]} : x^2$ relationship from linear form as being due to boundary diffusion and diffusion along dislocations.

The authors of [4] studied the relation between the self–diffusion velocity and the crystallographic direction in the α phase of uranium by absorption and autoradiographic methods. The investigation was carried out with single crystals, polycrystalline samples with large, perfect grains, and polycrystalline samples with coarse, imperfect grains at 640°C. The lowest value of self–diffusion coefficient corresponded to the [010] direction; approximate calculation gave a value of $D \leq 10^{-15}$ cm²/sec for this direction, and $D = 2 \times 10^{-13}$ cm²/sec for the [100] and [001]. The self–diffusion coefficients of uranium were also measured at 560 to 660°C in directions perpendicular to planes close to the (100) and (001). The temperature dependence of the self–diffusion coefficient of α–uranium along these directions is given by the following expression:

$$D = 9.1 \cdot 10^{-2} \cdot \exp(-67,000/RT) \text{ cm}^2/\text{sec.}$$

There is a tendency for the activation energy to fall as the angle between the direction in question and the [010] direction diminishes.

Thus the results of these various papers (shown graphically in Fig. 1) are, first, extremely contradictory, and secondly far from complete, since the temperature range studied is small and the variation of the boundary and volume diffusion coefficients in α–uranium with temperature is not given.

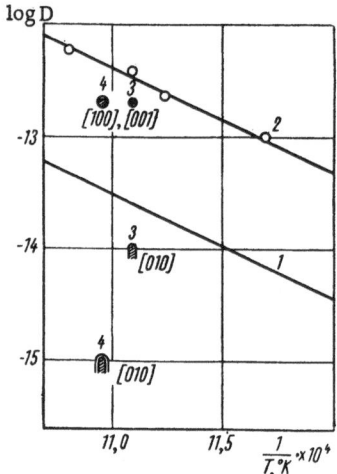

Fig. 1. Comparative data on the self-diffusion coefficients of α-uranium obtained by different authors. 1) Our own data (volume diffusion); 2) temperature dependence according to [2]; 3) self-diffusion coefficient at 625°C for various crystallographic directions [3]; 4) ditto for 640°C according to [4].

In the present investigation we therefore decided to determine the temperature dependence of the boundary and volume self-diffusion coefficients separately over a wider temperature range.

We used polycrystalline samples of electrolytic uranium (99.87% purity). The samples were prepared in the form of cylinders 10 to 12 mm in diameter and 6 to 8 mm long. The faces were made parallel and the ends of the samples were cleaned up by grinding with an emery-paper trimming device.

Before studying the diffusion in the samples, these were placed in sealed and evacuated quartz capsules, water-quenched from 800°C, and stabilized by annealing for 4 h at 630°C. A micrograph of the original structure of the uranium appears in Fig. 2.

Uranium enriched (90%) with the U^{235} isotope was used as radioactive indicator for studying self-diffusion. Chippings of the enriched uranium were suspended inside a tungsten heater and vacuum-evaporated on to the end surfaces of the samples. The samples so treated, having a surface activity of 10 to 13 × 10^3 pulses/min (natural uranium with the same surface has an activity of some 3 × 10^3 pulses/min), were tied together in pairs by means of tungsten wire, the active sides inward. Then the samples were placed in quartz capsules, which were evacuated to a pressure of 10^{-5} mm Hg and sealed. Diffusion annealing was carried out at 630, 590, 550, and 500°C for 434, 1455, 1827, and 1835 h, respectively. The furnace temperature was automatically held constant to ±5°C.

After diffusion annealing, the samples were taken out of the capsules and subjected to layer-by-layer radiometric analysis. A layer some 500 μ thick was first removed from the sides of the samples in order to eliminate the effects of surface diffusion. The layers were removed manually, using emery paper.

The depth of penetration of the active material into the samples was only 10 to 20 μ. The thickness of the layer removed was estimated directly, using a beam micrometer with an accuracy of ±1 μ, and also by measuring the thickness of several tens of layers. The relative error in measuring the thickness of the abraded layer was no greater than ± 10%.

The radioactivity of the remaining part of the sample was measured by an α particle scintillation counter.

In view of the fact that the path of α particles in uranium is approximately 3 μ, we may suppose that the difference between the total activity of the remaining part of the sample (as measured by the counter) and the background (the α activity of the natural uranium before depositing the U^{235} layer) is proportional to the specific radioactivity i and characterizes the concentration of the diffusing uranium isotopes.

Analysis of the results showed that in no case did the logarithm of the specific activity depend linearly on the square of the penetration depth. The log i : x^2 curves had a concave form, indicating the existence of boundary diffusion in the sample studied.

In our case the relation between the specific activity and the penetration depth had a linear character, which enabled the coefficients of boundary diffusion to be calculated by the method

TABLE 1. Coefficients and Parameters of Volume Self-Diffusion in α-Uranium

Coefficient of self-diffusion ($\times 10^{-15}$ cm^2/sec) at temp. (°C)				Q, kcal/g·atom	D, cm^2/sec
630	590	550	500		
25	9.4	2.7	0.48	42.2	$4.5 \cdot 10^{-4}$

TABLE 2. Coefficients and Parameters of Boundary Self-Diffusion in α-Uranium

Coefficient of self-diffusion ($\times 10^{-17}$ cm^2/sec) at temp. (°C)				Q, kcal/g·atom	D, cm^2/sec
630	590	550	500		
9.6	3.1	0.8	0.14	44.3	$1.6 \cdot 10^{-5}$

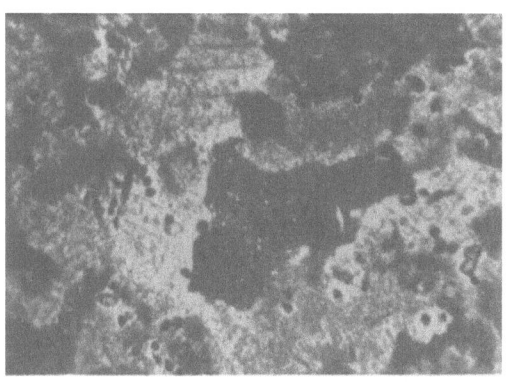

Fig. 2. Micrograph of the original structure of the uranium ($\times 300$).

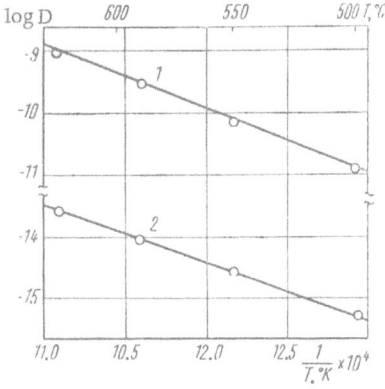

Fig. 3. Temperature dependence of the self-diffusion coefficients of α-uranium: 1) Boundary diffusion; 2) volume diffusion.

proposed by Fisher [5]. This method is based on certain simplifying assumptions and required independent determination of the coefficients of volume diffusion. In our investigation the latter were determined from the initial sections of the concentration curves, which we considered a very objective estimate for the penetration of the U^{235} atoms into the body of the grains in the polycrystalline samples. The same method of calculating the coefficients of volume diffusion was used in [3]. On the initial part of the curve there was a linear relationship between log i and x^2, and the diffusion coefficients were determined by Gruzin's method [6]. The coefficients and parameters of volume self-diffusion in α-uranium are given in Table 1. The temperature dependence of the self-diffusion coefficients are shown graphically in Fig. 3 (curve 2).

Calculation of the boundary coefficients by Fisher's method does not give the absolute value of D_b, since for this we must know the width of the grain boundary δ. Values calculated by using data on the coefficients of volume diffusion (see Table 1) are presented in Table 2.

In order to confirm the existence of boundary diffusion in α-uranium, the authors carried out layer-by-layer autoradiography of the samples. In taking the autoradiograms, film of the MK type was used. The holding time varied from one to four days.

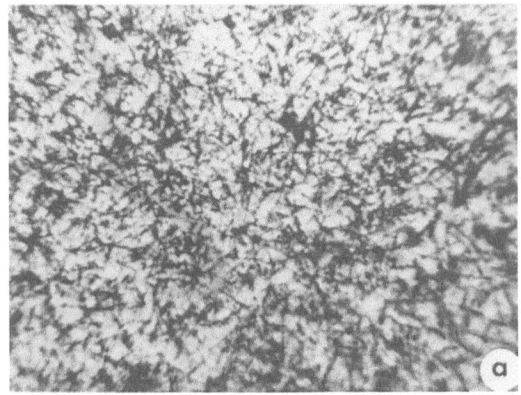

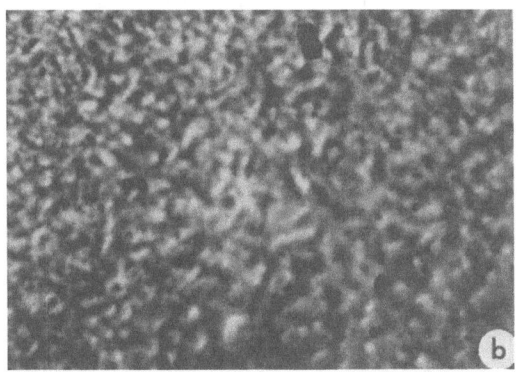

Fig. 4. Autoradiograms taken from a sample of pure uranium (annealed at 550°C) at a depth of 8 μ (×500): a) Focused image; 2) defocused image.

Figure 4 shows autoradiograms of a uranium sample annealed at 550°C after removing an 8-μ layer. We notice that α particle tracks are clearly visible on the focused pictures (Fig. 4a). If we look at a slightly defocused image under the microscope, however, we get the impression of preferential diffusion along the boundaries of integration blocks 3 to 5 μ in size. It should be noted that the grains themselves were certainly larger than this (see Fig. 2).

These results indicate that the activation energy of boundary diffusion in uranium differs only very slightly from the corresponding value for volume diffusion. At the same time, the level of diffusion mobility along the grain boundaries considerably exceeds volume diffusion.

By way of example, Fig. 3 compares the temperature dependence of the coefficients of boundary and volume diffusion in α-uranium (curves 1 and 2, respectively). The absolute values of the coefficients of boundary diffusion are given on the assumption that the width of the grain boundary $\delta = 10^{-7}$ cm. Under these conditions, the velocity of boundary diffusion exceeds that of volume diffusion by 4.5 orders of magnitude. Even if we take $\delta = 10^{-5}$ cm, as suggested by V. I. Arkharov [7], there will be a considerable difference between the coefficients.

It has been established that for the vacancy mechanism the self-diffusion activation energy may be represented as the sum of the energy of forming the vacancies ($^1/_3$ of the total energy) and the activation energy for the motion of vacancies ($^2/_3$ of the total) [8]. In addition, to this, it is known that the activation energy of boundary diffusion is approximately equal to the activation energy for the motion of the vacancies, since the boundaries are very much loosened and saturated with vacancies.

We may suppose that in the course of phase transformations in uranium the intergrain boundaries of the old phase may not "heal up" for a long time, since the process by which the old grains vanish is of a diffusion nature. In α-uranium the self-diffusion coefficients have an order of magnitude of 10^{-14} cm^2/sec, even in the high-temperature range, i.e., extremely small. For this region the old "unhealed" grain boundaries remain sources for the appearance of vacancies of α-uranium leads to the development of large microstresses in the course of thermal cancies. In addition to this the severe anisotropy in the properties of α-uranium leads to the development of large microstresses in the course of thermal variations, and this also may provide an additional source of vacancies.

All this helps to explain the approximate agreement between the activation energies of boundary and volume diffusion, since the latter is almost entirely the activation energy for the motion of vacancies.

Our own values for the activation energy of self-diffusion practically coincide with those given in [2] for polycrystalline α-uranium. If in accordance with the foregoing considerations we take 40 kcal/g $\cdot$ atom as the energy for the motion of the vacancies, the true activation energy of self-diffusion for α uranium should be around 60 kcal/g $\cdot$ atom, which agrees closely with the value given by the 40 T_m formula.

Our data on the volume diffusion coefficients lie between those of [1] and [3] obtained on single crystals of α-uranium.

Conclusions

1. At α phase temperatures, self-diffusion in uranium takes place mainly along the boundaries of intragrain blocks.

2. We have obtained the coefficients and parameters of the boundary and volume self-diffusion of uranium in the α phase. The level of diffusion mobility along the grain boundaries is considerably (four to five orders) greater than the volume-diffusion figure. The self-diffusion coefficients vary with temperature in the following way:

$$D_{vol} = 4.5 \cdot 10^{-4} \exp\left(-42{,}200/RT\right) \text{ cm}^2\text{/sec},$$
$$D_b\delta = 1.6 \cdot 10^{-5} \exp\left(-44{,}300/RT\right) \text{ cm}^3\text{/sec}.$$

Literature Cited

1. Resnik, R., et al. J. Nucl. Mater., 5:5 (1962).
2. Adda, Y., et al. Compt. Rend., 253:445 (1961).
3. Rothman, S. J., et al. J. Appl. Phys., 33:2113 (1962).
4. Bochvar, A. A., et al. Paper No. P/233 (USSR) presented to the Third International Conference on the Peaceful Use of Atomic Energy (Geneva, 1964).
5. Fisher, J. C. Appl. Phys., 22:74 (1951).
6. Gruzin, P. L. Probl. Metalloved. i Fiz. Metal., Inst. Metalloved. i Fiz. Metal., Sb. Tr., No. 3:201 (1952).
7. Arkharov, V. I., and Skornyakov, N. N. Tr. Inst. Fiz. Metal., Akad. Nauk SSSR, Ural. Filial, 16:75 (1955).
8. Gertsriken, S. D., and Dekhtyar, I. Ya. Diffusion in Metals and Alloys in the Solid Phase, Moscow, Fizmatgiz, 1960.

PHASE DIAGRAM OF THE UC-W SYSTEM AND DIFFUSION OF URANIUM FROM URANIUM CARBIDE INTO TUNGSTEN

A. I. Evstyukhin, G. B. Fedorov, G. I. Solov'ev, E. A. Smirnov, F. I. Zhomov, and A. G. Zaluzhnyi

Of considerable interest in atomic-power systems are fuel elements using refractory uranium-carbon compounds (uranium monocarbide and dicarbide [1-3]) as fuel. It is thus important to study the phase diagram and compatibility of uranium-monocarbide fuel with refractory materials.

We therefore decided to study the phase diagram of alloys in the UC−W system, to examine the compatibility of uranium monocarbide with tungsten, and to determine the quantitative characteristics of the diffusion of uranium from the monocarbide into tungsten.

Rudy et al. [4] plotted an isothermal section of the ternary U−W−C phase diagram at 1500°C, indicating the existence of a quasi-binary UC−W cross section. No details were given regarding this section.

We ourselves studied the UC−W phase diagram by determining the temperature at the onset of melting for these alloys and also by x-ray diffraction and metallographic analysis. Our alloys were prepared by briquetting powdered uranium carbide and tungsten at a pressure of about 5×10^3 kg/cm^2 and sintering in a furnace with a graphite heater at 2000°C and 1×10^{-4} mm Hg, subsequently remelting in an MIFI-9-3 arc furnace.

The original material for producing the uranium carbide was 99.87 wt.%-pure commercial uranium and spectroscopically pure graphite in the form of 5-mm-diameter rods. The UC was obtained by the arc melting.

Table 1 shows the compositions of the alloys studied.

The temperature of the onset of melting was determined by direct measurement with an OP-48 optical pyrometer (accuracy ±20°C). In order to eliminate errors associated with the radiation of a nonabsolute blackbody, the temperature measurement was referred to the bottom of a blind opening in the center of the sample with a diameter/depth ratio not greater than 1 : 5. The sample was heated by direct passage of a current in an atmosphere of purified helium.

The x-ray diffraction study of the alloys was based on the Debye method, using a cylindrical RKU-86 camera and filtered copper radiation.

TABLE 1. Composition of the Alloys

No. of alloy	Composition as per mixture, wt. %			Content (as per chemical analysis), wt. %	Temperature of the onset of melting, °C
	U	W	C		
1	94.25	1	4.75	4.62	2370
2	93.30	2	4.70	4.42	2310
3	91.39	4	4.61	4.49	2290
4	90.44	5	4.56	4.67	2280
5	85.68	10	4.32	4.68	2210
6	80.92	15	4.08	4.28	2260
7	76.16	20	3.84	3.64	2270
8	66.64	30	3.36	3.43	2110
9	57.12	40	2.88	2.96	2220
10	47.60	50	2.40	2.27	2140
11	38.08	60	1.92	1.82	2150
12	28.56	70	1.44	1.42	2140
13	19.05	80	0.95	0.91	2380
14	9.52	90	0.48	0.35	2400

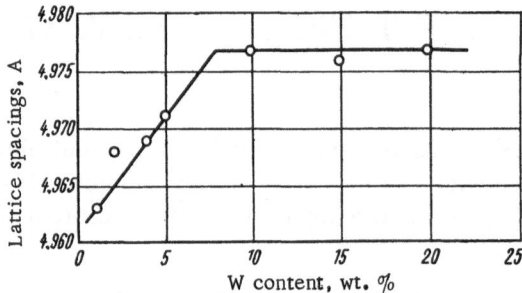

Fig. 1. Lattice spacing of cast UC−W alloys as a function of tungsten content in the monocarbide.

It was found that cast alloys No. 1-4 (Table 1) were single-phased. These had an fcc lattice, the parameter of which increased with increasing tungsten content. Figure 1 shows the variation in the lattice parameter of cast samples with changing tungsten content. The x-ray diffraction pictures of alloys No. 5-14 showed two series of lines, belonging to uranium monocarbide and tungsten. It was found that the maximum solubility of tungsten in the case samples was approximately 8 wt.%.

Alloy samples were also water-quenched from temperatures of 1800 and 2000°C.

On examining the quenched alloys by x-ray diffraction, two series of lines belonging to uranium monocarbide and tungsten, respectively, were also found.

The solubility of tungsten in uranium monocarbide was lower than the earlier value. At 2000°C it was about 4 wt.%.

Metallographic study of the cast and quenched alloys showed that these had both single- and double-phased structures. Cast alloys No. 1-4 had the structure of a solid solution, and alloys No. 5-14 had two structural components, one of which was a eutectic.

Figure 2 shows the microstructure of alloy No. 1, which was single-phased, and Fig. 3 shows alloy No. 6, containing the eutectic and tungsten crystals. Analogous structures occurred for the quenched alloys.

The information obtained from the melting-point measurements and from the x-ray and metallographic analyses of the alloys was used to construct the UC−W phase diagram presented

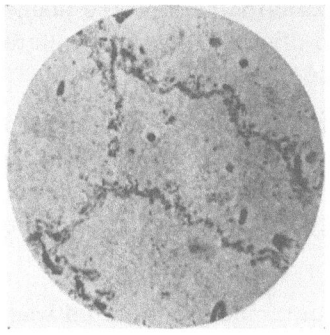

Fig. 2. Microstructure of a cast alloy containing 99 wt.% UC and 1 wt.% W (×335).

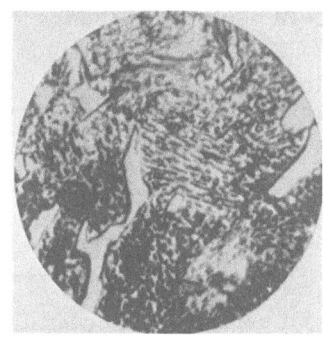

Fig. 3. Microstructure of a cast alloy containing 85 wt.% UC and 15 wt.% W (×335).

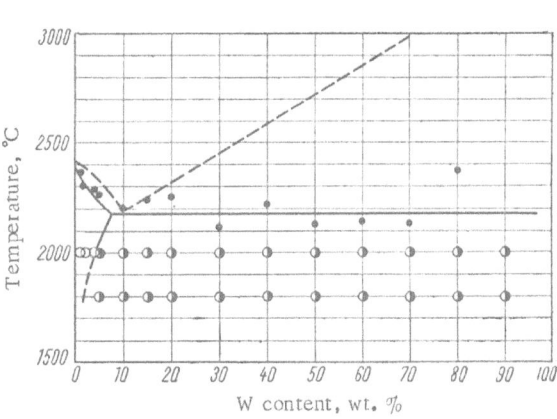

Fig. 4. Phase diagram of the UC−W system: ○) Single-phase alloys; ◑) two-phased alloys; ●) temperature at the onset of melting in the alloys.

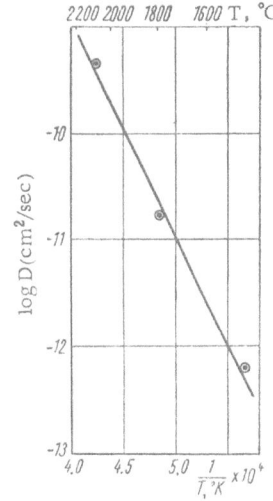

Fig. 5. Diffusion coefficients of uranium in tungsten as a function of temperature.

in Fig. 4. The UC−W phase diagram is a diagram of the eutectic type with limited solubility of tungsten in uranium monocarbide in the solid state. The melting point of the eutectic equals 2180 ± 20°C. Metallographic analysis indicated the existence of a eutectic containing approximately 10 wt.% tungsten. No new chemical compounds were found in the system.

Chemical analysis showed that the samples of uranium monocarbide remelted for the diffusion tests contained 4.9 ± 0.1 wt.% carbon (the stoichiometric composition of uranium monocarbide corresponds to 4.8 wt.% carbon). The tungsten samples were obtained from pressed and sintered rod material of 99.94% purity in the form of cylinders 8 to 10 mm in diameter and 7 to 12 mm long. In order to ensure that the ends of the samples should be parallel, the samples were finished with an emery-paper trimming device and a polishing machine.

The diffusion annealing was carried out in TVV–4 furnaces over the range 1500 to 2100°C. The samples were annealed in the form of diffusion pairs with uranium carbide placed between

TABLE 2. Coefficients of Diffusion of
Uranium from Uranium Monocarbide
into Tungsten

T, °C	D, cm²/sec
1495	$5.9 \cdot 10^{-13}$
1792	$1.7 \cdot 10^{-11}$
2096	$4.6 \cdot 10^{-10}$

them. The diffusion pairs were bound with tungsten wire and placed on tungsten plates in the working space of the furnace.

The temperature of the samples was measured with an OPPIR-09 optical pyrometer. Corrections were added to the temperatures measured to allow for the radiation of a nonabsolute blackbody.

The almost complete absence of absorption in the intermediate medium was checked by reference to the melting points of iron and zirconium. The absolute error in measuring the temperatures of sample annealing was no greater than ±20°C.

After diffusion annealing had ended, the initial activity of the samples was measured and then the diffusion layers were removed. Traces of uranium carbide were first removed from the ends and sides of the samples. The layers were removed manually with a polishing cloth. The penetration depth of the uranium into the tungsten was some 0.002 cm on average. The thickness of the layer removed was estimated directly with a micrometer of the lever type to an accuracy of ±0.0001 cm, and also by measuring the thickness of several removed layers (to 0.005 cm) with a micrometer. The relative error in measuring the thickness of a removed layer was no greater than ±10%.

The radioactivity of the remaining part of the samples was measured with a scintillation counter. The diffusion coefficients were calculated by the Grube [5] method, considering diffusion from a constant source.

The diffusion coefficients determined in this way are presented in Table 2.

The temperature dependence of the diffusion coefficients of uranium in tungsten has the following form:

$$D = 0.11 \exp(-91,700/RT) \text{ cm}^2/\text{sec.}$$

The experimental results are shown graphically in Fig. 5.

On annealing uranium monocarbide in contact with pure metals, we may expect the formation of a zone of interaction with the creation of new carbide phases based on these pure metals by the "drawing" of carbon from the uranium monocarbide. In our case no interaction of tungsten with the uranium monocarbide was observed. This entirely agrees with data relating to the free energies of formation for the carbide phases of the metals in question. The change in free energy for tungsten carbide is given by the equation [6]:

$$\Delta F = -9100 + 0.4T \text{ cal/g} \cdot \text{atom.}$$

For uranium monocarbide [7]

$$\Delta F = -22,440 + 13.13 \cdot T \log T - 2.46 \cdot 10^{-3} \cdot T^2 + 0.35 \cdot 10^5 \cdot T^{-1} - 33.5 \cdot T \text{ cal/g} \cdot \text{atom.}$$

The free energy for tungsten and uranium carbides calculated from these equations at 2073°K are −8300 and −15,000 cal/g ·atom, respectively. Thus the uranium carbide phase is the more stable in our temperature range; this prevents the interaction of uranium carbide with tungsten to form tungsten carbide.

Conclusions

1. We have plotted the phase diagram of the UC−W system. This diagram is of the eutectic type with a limited solubility in the solid state. The melting point of the eutectic is 2180 ± 20°C. The eutectic point is 10 wt.% W.

2. The solubility of tungsten at the melting point of the eutectic is around 8 wt.%.

3. The solubility of tungsten at 2000°C is around 4 wt.%.

4. The solubility of uranium monocarbide in tungsten is negligible.

5. We have studied the diffusion of uranium from uranium monocarbide into tungsten over the temperature range 1500 to 2100°C. The temperature dependence of the diffusion coefficient takes the form

$$D = 0.11 \cdot \exp\left(-91{,}700/RT\right) \text{ cm}^2/\text{sec}.$$

6. There is no interaction between the uranium carbide and tungsten over the temperature range studied.

Literature Cited

1. Zaimovskii, A. S., et al. Fuel Elements of Atomic Reactors, Moscow, Gosatomizdat, 1962.
2. Holden, A. N. Physical Metallurgy of Uranium [Russian translation], Moscow, Gosatomizdat, 1962. [English edition: Reading, Penn., Addison-Wesley Publishing Co., Inc., 1958.]
3. Materials for Nuclear Reactors, Moscow, Gosatomizdat, 1963.
4. Rudy, E. Mondshefte für Chemie, No. 2 (1962).
5. Grube, G., and Jedele, A. Z. Electrochem., 38:799 (1932).
6. Ackermann, R. J., and Thorn, R. J. International Conference on the Peaceful Uses of Atomic Energy, 1958, 2nd, Geneva, New York, Pergamon Press, Inc.
7. Strasser, A. Atomnaya Tekhnika Za Rubezhom, 1:24 (1961).

VARIATION OF HARDNESS DURING THE ISOCHRONAL AND ISOTHERMAL ANNEALING OF WORKED URANIUM WITH VARIOUS DEGREES OF PURITY

A. A. Tsvetaev, R. K. Chuzhko,
V. N. Bovenko, and Yu. N. Golovanov

It is well known that, owing to the poor solubility of impurities in α uranium [1], aging processes set in after irradiation and quenching [2, 3]. There is also a process of strain aging, reflected in the "humps" on the curves relating the mechanical properties of many metals to temperature [4]. There is an anomalous rise [5, 6] in the elastic limit, hardness, and elastic modulus in the region of recrystallization annealing for some metals.

In uranium these processes have been little studied, especially the latter two. We therefore decided to study the general picture of the recovery of hardness in the isochronal annealing of uranium of various degrees of purity after working, and also to determine the activation energy of the stress-relaxation process.

Method

We chose uranium of various degrees of purity for study. Table 1 shows the chemical analyses of the uranium samples.

The samples for study were prepared by extruding rods 8 mm in diameter. The rods were cut into small cylinders 6 mm long. After twofold phase recrystallization (to eliminate texture), the samples were annealed for 6 h at 600°C and then subjected to a 50% upsetting in a press at 0°C. Subsequently the samples were submitted to isochronal (Δt = 30 min) and isothermal annealing. The isochronal annealing enabled the general picture of the recovery of properties to be seen [7].

The hardness was measured after annealing with a Vickers hardness gage, the data from 8 to 10 measurements being averaged.

Results of Experiments

Figure 1 shows the change in the hardness (H_V) of uranium and uranium alloys on isochronal annealing. The first feature on these curves is the hardness peak at 100 to 140°C.

The higher the impurity content, the higher was the temperature corresponding to maximum hardness. The peak was sharpest for alloy No. 1.

TABLE 1. Impurity Content of the Uranium Studied
(in %)

Sample	Fe	Al	Si	C
Pure uranium . . .	0.002	0.001	0.001	0.05
Alloy No. 1 . . .	0.008	0.005	0.007	0.08
Alloy No. 2	0.022	0.005	0.046	0.07

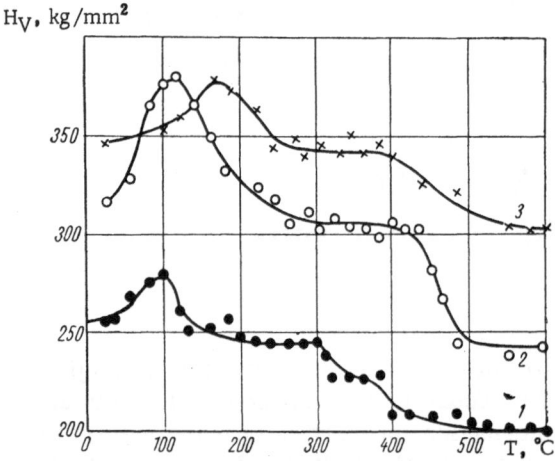

Fig. 1. Recovery of hardness during the iso-chronal annealing of worked uranium (an-nealing time 30 min): 1) Pure uranium; 2) alloy No. 1; 3) alloy No. 2.

Fig. 2. Variation in the degree of recovery (γ) for isochronal annealing: 1) Pure uranium; 2) alloy No. 1; 3) alloy No. 2.

As the annealing temperature rose, the hardness fell to its original level (in the de-formed or worked state) and then fell further, starting from a particular temperature for each alloy. Increasing the impurity content raised the temperature at which stress removal began.

Figure 2 shows the degree of recovery (expressed as the ratio of the hardness increment after annealing at T_i to the original increment) as a function of temperature. We see from these curves that the hardening effect is greatest for alloy No. 1; this is no doubt due to the op-timum combination of physical- and chemical-defect concentrations. Whereas, for pure urani-um and alloy No. 1, the degree of recovery tends to zero on annealing at the upper boundary of the α form of the metal, in alloy No. 2 it remains at some 50% of the original hardness incre-ment.

Discussion of Results

Strain Aging in Uranium. The homologous temperatures of the maxima on the curves of Figs. 1 and 2 are 0.27, 0.28, and 0.3 for pure uranium and alloys Nos. 1 and 2, re-spectively.

It is well known that the graphs relating the mechanical properties of many metals to temperature shows strain-aging peaks. The homologous temperatures (θ) of the peaks are shown in Table 2 [4] for a number of metals.

Thus the peaks observed lie precisely in the temperature range of strain aging. Aging after irradiation and quenching is also found in this temperature range [3]. The mechanism

TABLE 2. Homologous Temperatures of a Number
of Metals

Type of lattice	Metal	T, °C	θ
bcc	Mo Fe W	307 197 693	0.2 0.26 0.265
fcc	Ni Al	247 82	0.3 0.38
hcp	Ti Co	497 452	0.385 0.41
Orthorhombic	U	110	0.28 [8]

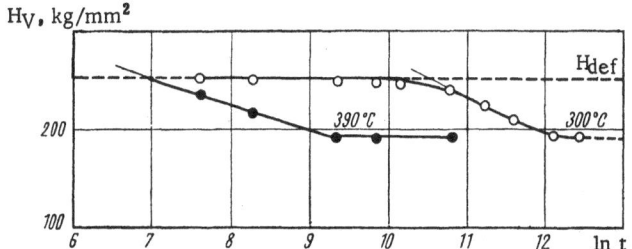

Fig. 3. Recovery of hardness for isothermal annealing.

underlying the hardness peaks in the isochronal annealing of worked uranium apparently lies
in the diffusion of substitution-type impurities into the elastically-distorted regions produced
by working. The diffusion of the impurities is facilitated (the temperatures of the peaks are
quite low) by the presence of vacancy concentrations in excess of equilibrium; in uranium these
become mobile in the temperature range 35 to 150°C [3].

Activation Energy of Self-Diffusion in Uranium. The relaxation proc-
esses occurring in metals are due to the recovery of the properties of the crystal lattice when
defects in excess of equilibrium are annihilated by the directional motion of vacancies. On this
principle, we may consider that the relaxation of elastic internal stresses should ultimately
be reflected in the development of residual (plastic) strains.

Assuming the vacancy mechanism of plastic strain (Nabarro [9]) during stress relaxation, we may [10] obtain the following law for the variation in internal stresses during isothermal annealing:

$$\sigma = \sigma_0 - \frac{kT}{v} \ln\left(\frac{t}{t_0} + 1\right),\tag{1}$$

where σ and σ_0 are the stresses at instants t and 0, k is the Boltzmann constant, T the absolute temperature, and v the volume corresponding to one atom.

The value of t_0 is determined in the following way:

$$t_0 = \frac{kTl^2}{2vED_0} \exp\left(\frac{U_0 - v\sigma_0}{kT}\right),\tag{2}$$

where U_0 is the self–diffusion activation energy, D_0 the pre–exponential factor in the expression for the diffusion coefficient, E the elastic modulus, and l the average diffusion distance of vacancies to the point of annihilation.

According to expression (1), the fall in stress for $t > t_0$ should be proportional to the logarithm of the annealing time. If we suppose that the hardness is proportional to the internal stresses, then the variation of hardness with $\ln t$ should be analogous. The form of the curves in Fig. 3 confirms this. Hence we shall suppose that the hardness H is proportional to the internal stresses σ_i, i.e.,

$$\sigma_i = \beta H, \tag{3}$$

where β is a proportionality factor. Then expressions (1) and (2) may be written in the form

$$H = H_0 - \frac{kT}{\beta v} \ln\left(\frac{t}{t_0} + 1\right), \tag{4}$$

$$t_0 = \frac{kTl^2}{2vED_0} \exp\left(\frac{U_0 - \beta v H_0}{kT}\right). \tag{5}$$

The value of βv may be determined from the slope of the straight lines in Fig. 3; it is in fact 3.8×10^{-23} cm^3.

On this basis, $\sigma_i \approx 2H$. According to experimental data, H_0 is 256 kg/mm^2; hence the initial microstresses may be approximately 500 kg/mm^2. If we consider that the macrostresses in deformed uranium may reach σ_s, i.e., around 20 kg/mm^2, the stress–concentration coefficient q approximately equals 25.

The self–diffusion activation energy was calculated from Eq. (3) on the basis of the experimental values of βv and t_0 at two temperatures, using the following formula:

$$U_0 = k \frac{T_1 T_2}{T_2 - T_1}\left(\ln \frac{t_0}{t_0''} + \ln \frac{T_2}{T_1} + \ln \frac{E_1}{E_2}\right) + \beta v H_0. \tag{6}$$

The values of E were taken from [8]; then the activation energy U_0 became 45,000 cal/mole, agreeing with published data [11].

Conclusions

1. The maximum on the isochronal hardness-recovery curve is due to strain-aging processes.

2. The recovery of hardness at $T > 300°C$ is due to processes of self–diffusion in the presence of internal stresses.

3. The activation energy of self–diffusion for α uranium is $U_0 = 45,000$ cal/mole.

Literature Cited

1.   Chemistry of Uranium [Russian translation], J. Katz and E. Rabinowicz (eds.), Moscow, IL, 1954. [English edition: Dover Publications, Inc., New York, 1951 (paper).]
2.   Holden, A. M. Physical Metallurgy of Uranium [Russian translation], Gosatomizdat, 1962. [English edition: Addison-Wesley Publishing Co., Inc., Reading, Penn., 1958.]
3.   Metallurgy of Reactor Materials. Reviews of the Battelle Institute. Book 1. Nuclear Fuel Materials [Russian translation], Skorov (ed.), Moscow, Gosatomizdat, 1961.
4.   Sokolov, L. D. Resistance of Metals to Plastic Deformation, Moscow, Metallurgizdat, 1963.

5. Rakhshtadt, A. G., et al. Metalloved. i Term. Obrabotka Metal., No. 1:45 (1962).

6. Spektor, É. N., et al. Fiz. Metal. i Metalloved., 17:445 (1964).

7. Mechan, C. J., and Brinkman, J. A. Phys. Rev., 103:1193 (1956).

8. Sergeev, G. Ya., et al. Physical Metallurgy of Uranium and Some Other Reactor Materials, Moscow, Atomizdat, 1960.

9. Nabarro, F. K. Phys. Soc., 75:(1948).

10. Platonov, P. A. Effects of Nuclear Radiations of Matter, Moscow, Izd. Akad. Nauk SSSR, 1962.

11. Adda, J., et al. Compt. Rend., 253:445 (1961).

CHANGE IN THE SHAPE OF METALS UNDER THE SIMULTANEOUS EFFECTS OF THERMAL CYCLING AND AN EXTERNAL LOAD

Yu. N. Golovanov, A. A. Tsvetaev, K. N. Gedgovd, and R. K. Chuzhko

There is a fairly well-developed phenomenological theory relating to changes in the shapes of bodies subjected to thermal cycling as a result of the relaxation of thermal stresses [1]. An attempt has been made to describe this phenomenon on the basis of the thermodynamics of irreversible processes [1]. The microscopic picture of the processes taking place during changes in shape, however, is still not clear, despite the large number of more or less successful mechanisms proposed by various authors [1-4]. The changes in the shapes of bodies resulting from thermal cycling may to some extent be considered as the result of a directional motion of crystal-lattice defects [2-5]. A factor orienting the motion of defects (in addition to texture) is the field of stresses developing in the body in the course of thermal cycling. Hence elucidating the relation between the stressed state and changes in the dimensions of a body subjected to thermal cycling constitutes an extremely interesting problem, the solution of which should explain the microscopic picture of shape changes.

In the present investigation we studied changes in the dimensions of zinc and aluminum samples during thermal cycling in the presence of an external load. For the zinc, in which changes in shape could be caused by the presence of microstructural stresses associated with thermal anisotropy, the thermal-cycling conditions were so chosen as not to produce substantial thermal stresses associated with thermal gradients. For aluminum, in which thermal-anisotropy stresses were absent more severe conditions, giving rise to thermal stresses, were chosen.

There have been a number of papers on the effects of thermal cycling on creep [6-12]. These papers, however, have mainly treated the effects of medium and large tensile loads, and the regions of small external loads has been touched upon very lightly. The effects of texture under these conditions have also been little studied. Our aim was therefore to extend this sphere of knowledge.

Experimental Method

We studied 99.99% pure zinc and aluminum. The metal was cast into spherical copper molds and then forged from sphere to cube at 100°C. According to [13] this kind if treatment creates an untextured state. Material for plane and cylindrical samples was obtained from

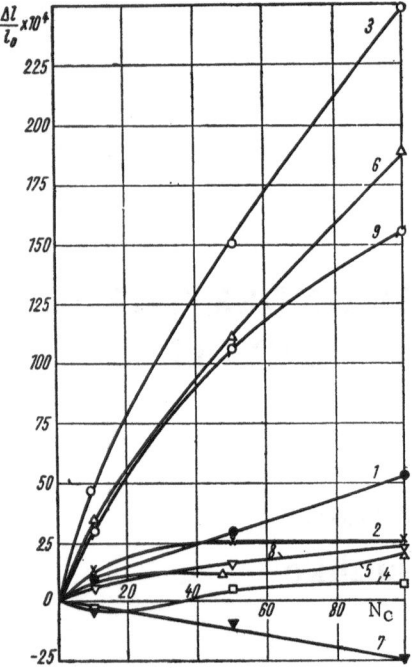

Fig. 1. Effect of an external tensile stress on the deformation of plane zinc samples: 1) After thermal cycling (samples forged and rolled, degree of reduction in rolling $\sigma = 80\%$); 2) after creep tests (same samples); 3) after thermal cycling under a load of $P = 100$ g/mm^2 (same samples); 4) after thermal cycling (samples forged); 5) after creep tests (samples forged); 6) after thermal cycling under a load of $P = 100$ g/mm^2; 7) after thermal cycling (samples forged and rolled, degree of rolling reduction $\sigma = 10\%$); 8) after creep tests (samples same); 9) after thermal cycling under a load of $P = 100$ g/mm^2 (same samples).

the texture-free metal by forging and rolling with various degrees of reduction. The plane samples were stamped out in a press and had a working part 25 mm long, 3 mm wide, and 1 mm thick, which gave a relatively uniform texture over the cross section [14]. The dimensions of the cylindrical samples were 3.5 and 4.8 mm in diameter and 15 mm in length.

Before thermal cycling, the zinc samples were annealed at 110 to 115°C for 8 to 10 h in oil; the aluminum samples were annealed for 4 h at 350°C in a muffle furnace. The thermal cycling was carried out in specially-constructed apparatus with automatic time and temperature control. The samples were heated in cylindrical furnaces in air. The maximum temperature of the cycle for zinc was $T_{max} = 240°C$ and the minimum $T_{min} = 50°C$ (air cooling). The heating time was 10 min for the plane samples and 18 min for the cylindrical; the cooling times were 10 and 12 min, respectively. The corresponding parameters for the aluminum samples were $T_{max} = 300°C$, $T_{min} = 10°C$ (water cooling). The heating time was 18 min and the holding time at the maximum and minimum temperatures 2 min. The accuracy of temperature regulation was ±2°C. The plane samples were tested under tensile loads applied by means of a system of weights. The cylindrical samples were tested under compressive loads created by calibrated 65G steel springs in special clamps. The calibration of the springs was checked after each experiment, and corrections were introduced when necessary.

The changes in the maximum dimensions of the samples between lines drawn at a certain distance from the ends were determined to an accuracy of ±0.005 mm with an IZA-2 comparator. The thickness of the plane samples was measured with a micrometer. In order to allow for the effect of changes in sample dimensions resulting from creep, parallel experiments were made with corresponding loads; in these the holding periods at the various temperatures corresponded to the holding periods in the course of thermal cycling.

Discussion of Experimental Results

Effect of an External Tensile Load.

Figure 1 shows the relative elongation of plane zinc samples ($\Delta l/l_0$) as a function of the number of cycles (N_c) after thermal cycling and after thermal cycling under the influence of a tensile load. For comparison the same figure shows the constant-temperature creep curves. The imposition of a tensile load leads to a considerable (several times) increase in the deformation as compared with that associated with sample thermal cycling or simple creep, or indeed the sum of these. Thus if the direction of the external load coincides with the direction of deformation during thermal cycling (curve 1), the deformation is increased many times (curve 3). Analogous results were obtained in [6-9, 12].

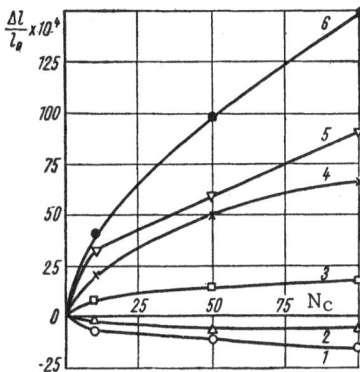

Fig. 2. Effect of an external tensile stress on the deformation of plane zinc samples during thermal cycling: 1) P = 0 (pure thermal cycling); 2) P = 3 g/mm^2; 3) P = 7 g/mm^2; 4) P = 15 g/mm^2; 5) P = 30 g/mm^2; 6) P = 100 g/mm^2.

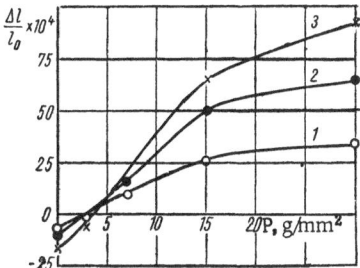

Fig. 3. Effect of the magnitude of the external tensile stress on the deformation of plane zinc samples during thermal cycling (1, 2, 3 indicate deformation after 10, 50, and 100 cycles, respectively).

If the deformation in the course of thermal cycling is close to zero (curve 4) or even negative (curve 7), the superposition of a tensile load leads to a positive deformation, considerably exceeding that associated with creep (curves 6 and 9). Thus the effects of the thermal cycling and the external load are not simply additive, but constitute a qualitatively new phenomenon.

Effect of the Load on the Deformation during Thermal Cycling. It is especially interesting to study the effect of changes in the external load on the deformation associated with thermal cycling in the low-load range (1 to 3 g/mm^2 for zinc), since this region has been little studied.

Figure 2 shows the deformation of plane zinc samples as a function of the number of thermal cycles under the influence of tensile loads of various magnitudes. In the course of 100-cycle tests these curves had a transient character and were very reminiscent of the initial (nonstationary)* part of creep curves at constant temperature.

In studying the effect of external loading on deformation during thermal-cycling tests, it is of particular interest to determine the magnitude of the external stresses which will suppress dimensional changes on thermal cycling, since this will to some extent indicate the value of the stresses producing deformation in thermal-cycling tests. For this purpose it is better to plot the curves of Fig. 2 again in coordinates of deformation against external stress. Such curves are shown in Fig. 3. The curves cut the x axis in the neighborhood of 3 to 3.5 g/mm^2. Thus for zinc samples subjected to thermal cycling and undergoing moderate deformation ($\Delta l / l_0 = 25 \times 10^{-4}$ in 100 cycles) it is sufficient to apply a tensile stress of approximately 3 g/mm^2 in order to suppress the change in dimensions completely.

In addition to the tensile stresses, we studied the influence of small (15 to 20 g/mm^2) compressive stresses. Curves for zinc are shown in Figs. 4 and 5. These curves are reminiscent of the corresponding relationships for tensile loads. On determining the degree of external compressive stresses suppressing the increase in the length of the sample during thermal-cycling tests, we find approximately 2 g/mm^2 for samples having an increment $\Delta l / l_0$ of 26×10^{-4} in 100 cycles.

Figure 6 shows deformation curves for ordinary thermal cycling, creep, and thermal cycling under a compressive load of P = 300 g/mm^2 for cylindrical aluminum samples. Despite the fact that ordinary thermal cycling increases the length of the samples, thermal cycling under load produces a more considerable shortening of the samples than that associated with creep.

_____

*Also called "primary" or "transient" stage of creep; the creep rate decreases with time-Tr.

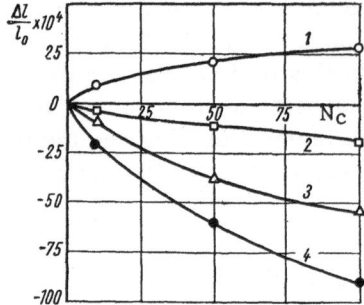

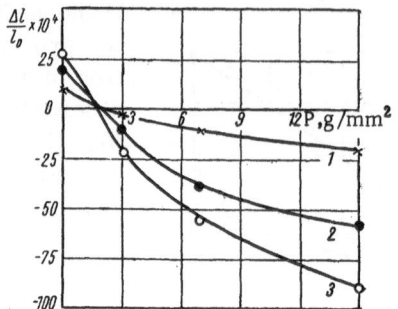

Fig. 4. Effect of an external compressive load on the deformation of cylindrical zinc samples during thermal cycling: 1) $P = 0$; 2) $P = 3$ g/mm^2; 3) $P = 7$ g/mm^2; 4) $P = 15$ g/mm^2.

Fig. 5. Effect of an external compressive load on the deformation of cylindrical zinc samples on thermal cycling (1, 2, 3 indicate deformation after 10, 50, and 100 cycles, respectively).

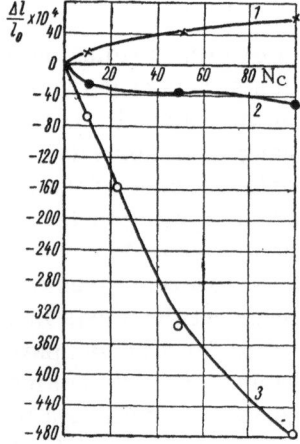

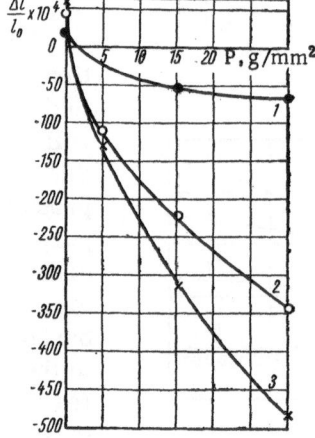

Fig. 6. Effect of an external compressive load on the deformation of cylindrical aluminum samples during thermal cycling: 1) After ordinary thermal cycling; 2) after creep tests ($P = 300$ g/mm^2); 3) after thermal cycling under load ($P = 300$ g/mm^2).

Fig. 7. Effect of the magnitude of the external compressive load on the deformation of cylindrical aluminum samples (1, 2, 3 indicate deformation after 10, 50, and 100 cycles respectively).

A study of the deformation as a function of compressive load enables us to determine the level of stresses compensating the increase in dimensions produced by thermal cycling. For aluminum samples having an increment $\Delta l / l_0 = 60 \times 10^{-4}$ in 100 cycles, this is approximately 10 to 12 g/mm^2 (Fig. 7).

Effect of Texture

Under thermal-cycling conditions, texture has an extremely substantial influence on the extent and sign of the deformation [1]. We have no reason to think that this will not also be so

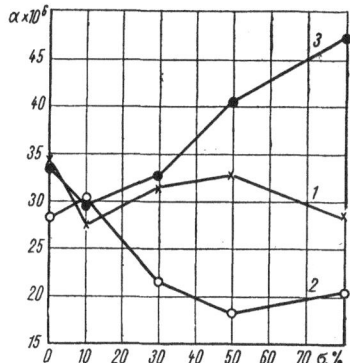

Fig. 8. Variation in the co-efficients of linear expansion (α) of plane zinc samples as a function of the degree of rolling reduction (σ, %): 1) α_x along the rolling direction (sample length); 2) α_y across the rolling direction (width); 3) α_z across the rolling direction (thickness).

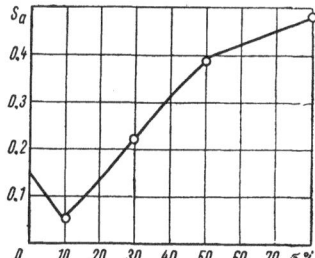

Fig. 9. Variation in the degree of anisotropy S_a on rolling plane zinc samples.

under load; hence it is interesting to compare the behavior of samples with different texture subjected to thermal cycling under an external stress. A series of plane samples was therefore prepared from cast zinc after first forging to remove texture and then rolling with different degrees of reduction (10 to 80%). The grain size after annealing was 0.08 to 0.2 mm). In order to characterize the texture, the coefficients of linear expansion were measured in three directions with a vacuum microdilatometer, one of these being the rolling direction (α_x along the sample) and the others being perpendicular to this (α_y over the width and α_z over the thickness). The changes in α_x, α_y, and α_z are shown as functions of reduction in Fig. 8.

In order to estimate the degree of anisotropy, we use the mean square deviation of α_i from the value α_p characterizing the sample in the detextured state and normalize this with respect to the mean square deviation of the coefficients $\alpha_\perp$ and $\alpha_\parallel$ of the hexagonal axis in a zinc single crystal from α_p. Then the quantity S_a, equal to

$$S_a = \sqrt{\frac{(\alpha_p - \alpha_x)^2 + (\alpha_p - \alpha_y)^2 + (\alpha_p - \alpha_z)^2}{2(\alpha_p - \alpha_\perp)^2 + (\alpha_p - \alpha_\parallel)^2}}, \qquad (1)$$

will characterize the degree of anisotropy in the plane sample. The variation of S_a with degree of rolling reduction is shown in Fig. 9. We see that in the original state the samples had a certain residual texture (probably after forging). Rolling with low degrees of reduction (around 10%) reduces the anisotropy considerably; further increasing the degree of reduction makes the anisotropy larger again.

Figure 10 shows the elongation of plane zinc samples during thermal cycling (ordinary and under load) as a function of the degree of rolling reduction. Comparing the elongation for thermal cycling (curve 1) with the curve $S_a = f(\sigma)$ (see Fig. 9), we notice that the elongation of the samples increases as texture intensifies, both curves having a transient character. It should be noted, however, that S_a only characterizes the degree of anisotropy and does not indicate the type of texture; hence it is not sufficient by itself to explain the effect of texture on deformation during thermal-cycling tests with or without load.

The imposition of an external tensile load of 100 g/mm² does not alter the way in which texture affects the deformation of samples under thermal cycling; it does however shift the whole curve in the direction of positive deformations (see Fig. 10, curve 2).

Figure 11 shows deformation as a function of external compressive load for two samples with different textures. Since in this case we were testing cylindrical zinc samples with axial symmetry, the texture was estimated from the so-called coefficient of anisotropy k_α, equal to the ratio of the linear-expansion coefficient along the axis of the sample coinciding with rolling direction ($\alpha_\parallel$) to that measured in a perpendicular direction ($\alpha_\perp$). Naturally $k_\alpha = 1$ for the detextured state. We see from Fig. 11 that in the sample with the lower degree of texture ($k_\alpha = 0.90$ as compared with $k_\alpha = 0.65$ for the other sample) the deformation in the direction of the external stress is greater (for the same magnitude of external load) and the external load re-

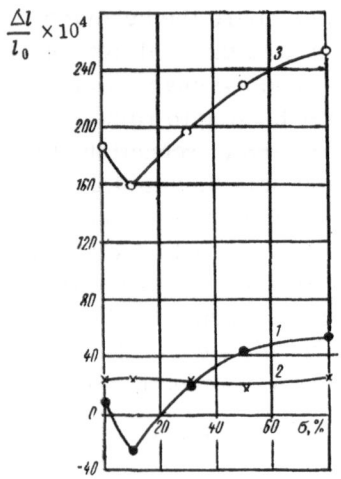

Fig. 10. Effect of the degree of rolling reduction on the deformation of plane zinc samples during thermal-cycling tests with and without an external tensile load and during creep tests: 1) Ordinary thermal cycling; 2) creep (P = 100 g/mm²); 3) thermal cycling under a tensile load (P = 100 g/mm²).

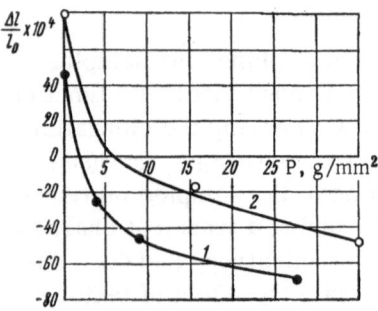

Fig. 11. Effect of the compressive load on the deformation of zinc samples: 1) Anisotropy coefficient $k_\alpha = 0.65$; 2) $k_\alpha = 0.90$.

quired to suppress the deformation is smaller. This is probably associated with the difference between the resultant thermal-anisotropy stresses in samples having different textures.

Discussion

Our own results and those of other authors [1, 6-12] show that thermal cycling under an external load produces an effect of dimensional instability considerably exceeding (by 5 or 10 times) the deformation resulting from ordinary thermal cycling or creep per se. This effect occurs both for metals with cubic lattices (aluminum) and metals with lower symmetry (zinc, uranium). A phenomenological picture of this behavior was presented in [1]. The authors of [2, 3] explain the deformation of wire samples and foil made from metals with cubic lattices (gold, silver, aluminum, etc.) by the quenching of vacancies with their subsequent annihilation at dislocation outcrops and grain boundaries. If we accept the vacancy mechanism of deformation caused by thermal cycling and cycling under load,* we can estimate the probable extent of the deformation resulting from the directional migration of point defects to sinks and compare it with the effect actually observed. It is well known that in addition to quenched vacancies thermal cycling may yield point defects in excess of equilibrium, these arising from internal stresses [16,17]. Internal stresses may arise either from quenching or as a result of the anisotropy of the linear-expansion coefficients, etc. According to [18], for some metals thermal-anisotropy stresses may reach very high values (for example, 40 to 125 g/mm² · deg for zinc).

It is known [19] that the concentration of equilibrium vacancies at temperature T may be calculated from the formula

$$\frac{n}{N} = Ae^{-\frac{U_V}{RT}}, \qquad (2)$$

where n is the number of vacancies, N is the number of atoms, U_V is the vacancy-formation energy, R the gas constant, and A = exp ($\Delta S/k$) is an entropy factor not depending on the temperature.

The results of calculating equilibrium vacancy concentrations from formula (2) are given in Table 1 (C = equilibrium concentration of vacancies at the quenching temperature; C_g = concentration of vacancies resulting from stress relaxation). For zinc T = 513°K, and according to [20] $U_V = 10,200$ cal/g-atom and A = 5.18. For aluminum T = 573°K.

Various values for the energy of vacancy formation appear in the literature: 11,200 to 12,900 cal/g-atom [20], approximately 14,000 cal/g-atom [21], 17,500 cal/g-atom [22, 23] (in

---

*While this article was being prepared, a new paper by V. A. Likhachev [15] using this approach appeared.

TABLE 1. Calculated Equilibrium Concentrations of Vacancies

Metal	C, $\times 10^{-4}$	C_g, $\times 10^{-4}$	$\Sigma(C + C_g)$, $\times 10^{-4}$	Deformation per cycle* for ordinary thermal cycling, $\varepsilon_{t.c.} \times 10^{-5}$	Deformation per cycle for creep*, $\varepsilon_c \times 10^{-5}$	Deformation per cycle for thermal cycling under load $\varepsilon_{t.c.+l} \times 10^{-4}$
Al	1	2	3	6	5	4
Zn	2.5	1†	5	5	2	2

*Our own data.

†Approximate calculation from the results of [16-18].

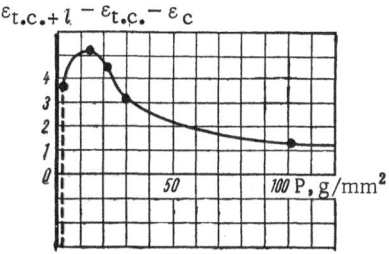

Fig. 12. Increment in the deformation of plane zinc samples during thermal cycling under a tensile load as a function of the magnitude of the load.

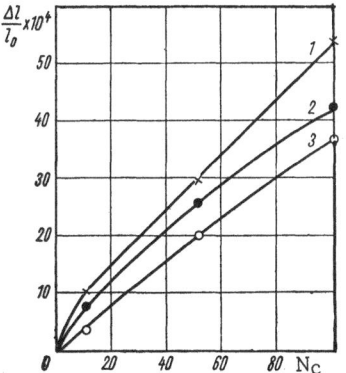

Fig. 13. Effect of thermal cycling on the elongation of zinc samples (σ = rolling reduction = 80%): 1) Samples cut parallel to the rolling direction; 2) samples cut at 45° to the rolling direction; 3) samples cut perpendicular to the rolling direction.

the view of the authors of [20], the last two values are too high on account of the possible influence of vacancy-complex formation). Hence for calculation purposes it seems fair to use an average value of U_V = 13,000 cal/g-atom and A = 10, in accordance with [19].

For comparison, Table 1 shows the concentrations of vacancies formed in the course of internal-stress relaxation and the orders of magnitude of the experimentally obtained deformations resulting from thermal cycling, creep tests, and thermal cycling under load. The number of vacancies formed on deformation is estimated by analogy with the estimate for gold in [17]. By calculating the quenching stresses for aluminum corresponding to a quenching speed of 800 deg/sec (recorded on a loop oscillograph for water quenching) from the data of [16], we obtained a value of 1.5 kg/mm². The quenching stresses in the zinc samples (heated and cooled quite slowly) were smallish, but as indicated earlier the thermal-anisotropy stresses reached very high values. All this once more indicates the existence of considerable deformation defects.

It thus follows from the data of Table 1 that in the metals studied the total number of defects arising is sufficient to give a plastic deformation close to that produced by thermal cycling under load, assuming a directional flow of all the defects present in excess equilibrium.

It is well known [24] that stress changes the shape of the potential barrier, facilitating the motion of atoms toward the lowered side of this. Hence the dimensional instability associated with thermal cycling can only occur if the stresses arising during the cycling process give asymmetric contributions to the directed flow of the surplus vacancies toward sinks.

Suppression of the thermal-cycling effect under given experimental conditions is achieved by applying a tensile stress to plane zinc and cylindrical aluminum samples (P ≈ 3 and 12 g/mm², respectively). Increasing the external load leads to a rise in the deformation effect, but this rise

TABLE 2. Relation between Texture Formation and Rolling Reduction

Degree of rolling reduction, %	$\varepsilon_0 = (\varepsilon_{t.c.+l} - \varepsilon_{t.c.} - \varepsilon_c) \times 10^4$
0	154
10	160
30	154
50	165
80	175

has a transient character for the range of applied stressed studied. Figure 12 shows the rise in the relative deformation produced by the specific action of the load during thermal cycling (referred to 1 g load) as a function of the tensile stress for plane zinc samples. The transient character of the curve indicates that defects capable of taking part in directional migration are vanishing. Thus we may say that the stress applied during thermal cycling intensifies the directional properties of the process whereby the surplus vacancies vanish. We may also suppose that the defects vanishing in the course of cooling are also annihilated in a directional manner. From this point of view the possible magnitude of the effect should increase on account of those defects which are annihilated chaotically on ordinary thermal cycling without any load. The existence of texture in anisotropic metals by itself gives a directional quality to the motion of vacancies [25]. On subjecting plane zinc samples (rolling reduction approximately 80%), cut along, across, and at 45° to the rolling direction, to thermal cycling with small heating and cooling rates, it was found that the length and breadth of the samples increased and the thickness diminished (Fig. 13). This kind of deformation is connected with the specific effects of the texture and thermal-anisotropy stress field, since it is well known that, on rolling zinc, the grain is oriented with the hexagonal axis at approximately 20 to 25° to the rolling direction [14], and since the vacancies move preferentially in the basal plane (owing to the low self-diffusion energy [25]) this leads to a reduction in the thickness of plane samples.

The deformation for thermal cycling under load $\varepsilon_{t.c.+l}$ may be put in the form

$$\varepsilon_{t.c.+l} = \varepsilon_{t.c.} + \varepsilon_c + \varepsilon_0 \tag{3}$$

where ε_0 is the deformation associated with the specific effect of the external load during the thermal cycling.

We see from Figs. 9 and 10 that a rise in the degree of anisotropy leads to a rise in $\varepsilon_{t.c.}$ while the value of ε_c hardly changes at all with changing texture. At the same time the data of Table 2 show that $\varepsilon_0 = \varepsilon_{t.c.+l} - \varepsilon_{t.c.} - \varepsilon_c$ also depends little on texture.

It thus follows that the main texture-sensitive term in Eq. (3) is $\varepsilon_{t.c.}$.

Thus in general the deformation arising from thermal cycling under an external load may be regarded as the result of the directional anisotropic motion of defects in the field of an asymmetric force constituting the resultant of interaction between internal and external stress fields.

Conclusions

1. The presence of external forces during thermal cycling greatly affects the deformation of the samples.

2. We have determined the magnitude of the external stresses which, on application to samples subjected to thermal cycling, exactly cancel the deformation caused by the same conditions of thermal cycling in the absence of external stress.

3. The texture-sensitivity of the deformation produced by thermal cycling is much the same whether there is an external load or not.

4. The deformation taking place in the course of thermal cycling is largely due to the directional displacement of surplus point defects (defects in excess of equilibrium) under the influence of internal stresses and texture.

5. The increase in deformation on applying an external load to the samples undergoing thermal cycling may be explained by supposing that the directional motion of the surplus point defects is intensified in the external stress field, both at the moment of quenching and in the course of subsequent heating and cooling.

Literature Cited

1. Davidenkov, N. N., and Likhachev, V. A. Irreversible Deformation of Metals during Thermal Cycling, Moscow-Leningrad, Mashgiz, 1962.
2. Dekhtyar, I. Ya., and Madatova, É. G. Fiz. Metall. i Metalloved., 6:939 (1958).
3. Dekhtyar, I.Ya., and Madatova, É. G. Vopr. Fiz. Metal. i Metalloved., Akad. Nauk Ukr. SSR, Sb. Nauchn. Rabot, No. 9:162 (1959).
4. Chiswick, Kelman. In book: Metallurgy of Nuclear Power and the Effects of Irradiation on Materials [Russian translation], Moscow, Gostekhizdat, 1956, pp. 612-641 (Contributions of non-Soviet scientists to the Internal Conference on the Peaceful Use of Atomic Energy, Geneva, 1955).
5. Likhachev, V. A., and Malygin, G. A. Fiz. Metal. i Metalloved., 16:435 (1963).
6. Likhachev, V. A., et al. Fiz. Metal. i Metalloved., 16(6):908 (1963).
7. Likhachev, V. A., and Malygin, G. A. Fiz. Metal. i Metalloved., 16:686 (1963).
8. Bochvar, A. A., et al. Atomnaya Énerg., 8(2):112 (1960).
9. Bochvar, A. A., et al. Issled. po Zharoproch. Splavam, Akad. Nauk SSSR, Inst. Met., 7:3 (1961).
10. McIntosh and Hill. In book: Transactions of the Second International Conference on the Peaceful Uses of Atomic Energy. Geneva, 1958. Selected contributions of non-Soviet scientists [Russian translation], Vol. 6. Moscow, Atomizdat, 1959, p. 187.
11. Young, A. G., et al. J. Nucl. Mater., 2:234 (1960).
12. Zalivadnyi, S. Ya., et al. Fiz. Metal. i Metalloved., 15(1):(1963).
13. Balakhovskii, O. A., et al. Zavodsk. Lab., No. 10:1207 (1962).
14. Barret, C. S. Structure of Metals [Russian translation], Moscow, Metallurgizdat, 1948. [English edition: McGraw-Hill, New York, 2nd ed.]
15. Likhachev, V. A., and Vladimirov, V. I. Fiz. Metal. i Metalloved., 17:665 (1964).
16. Van Bueren, H. G. Defects in Crystals [Russian translation], Moscow, IL, 1962.
17. Takamura, J. Acta Met., 9:(1961).
18. Likhachev, V. A. Fiz. Tverdogo Tela, 3(6):(1961).
19. Lomer, C. M. Usp. Fiz. Metal., 5:299 (1963).
20. Gertzriken, S. D., and Slyusar, B. F. Fiz. Metal. i Metalloved., 6:1061 (1958).
21. Nenko, B. S., and Kaufman, J. W. J. Phys. Soc. Japan, 15:(1960).
22. Bradshaw, F. J., and Pearson, S. Phil. Mag., 2:16, 570 (1957).
23. Simmons, R. O., and Baluffi, R. W. Phys. Rev., 117:52 (1960).
24. Frenkel, Ya. I. Introduction to the Theory of Metals, Moscow, Fizmatgiz, 1958.
25. Seith, W. Diffusion in Metals [Russian translation], Moscow, IL, 1958.

ABSORPTION METHOD OF DETERMINING THE DIFFUSION COEFFICIENT OF ALPHA-RADIOACTIVE ELEMENTS

L. V. Pavlinov and A. I. Nakonechnikov

It is well known that natural radioactive elements such as uranium, thorium, radium, etc., emit α particles in the course of decay. The existence of the α emission enables us to study the diffusion of these elements in metals and alloys. The short penetration distance of α particles in metals (a few microns) makes the absorption method of determining diffusion coefficients more convenient. This method was first used for studying self-diffusion in metallic lead and certain salts ($PbCl_2$ and PbI_2) using the radioactive lead isotope ThB [1-5]. The ThB was deposited on samples of the materials in question in the form of a thin layer. The diffusion coefficient was found from the well-known diffusion equation

$$c(x, t) = \frac{c_0 h}{\sqrt{\pi D t}} \exp\left(-\frac{x^2}{4Dt}\right),$$ (1)

where $c(x, t)$ is the concentration of the diffusing substance at a distance x from the surface after holding at a given temperature for a time t, c_0 is the initial concentration in the surface layer, h is the thickness of the layer, and D is the diffusion coefficient.

The concentration of the diffusing substance may be expressed quite simply in terms of the experimentally-measured value of the integral α activity of the surface layer on the sample before and after annealing. For the condition h < d (d is the layer corresponding to complete absorption of the α radiation) the initial value of the integral activity I_0 is proportional to the amount of material in the layer h, i.e.,

$$I_0 \sim c_0 h.$$ (2)

The integral α activity I of the surface layer of the sample after diffusion annealing is due to the radiation of atoms lying in a layer of thickness a. Hence

$$I \sim \int_0^a c(x)\,dx.$$ (3)

Taking account of (1) and (2) and defining the probability of the escape of an α particle by the coefficient $[1 - (x/a)]$, we obtain a final equation for calculating the diffusion coefficient in the form

$$\frac{I}{I_0} = \frac{2}{\sqrt{\pi}} \int_0^{\frac{\xi}{2}} e^{-u^2}\,du - \frac{1}{\xi\sqrt{\pi}}(1 - e^{-\xi^2}),$$ (4)

101

where

$$\xi = \frac{a}{2\sqrt{Dt}} .$$

(5)

Equation (4) is solved graphically and ξ is determined from the result. Relation (5) is then used to evaluate diffusion coefficient D.

Owing to the comparative complexity of the calculation, however, the absorption method has not been widely used. Recent data on the diffusion of uranium [6-8] were principally obtained by methods based on breaking up the sample.

The aim of our own investigation was to design a simple absorption method for finding the diffusion coefficient of α active elements. For calculating the diffusion coefficient we used Eq. (1). Putting (1) into (3) and taking account of (2), we write

$$I = \frac{I_0}{\sqrt{\pi Dt}} \int_0^a e^{-\frac{x^2}{4Dt}} dx .$$

(6)

In order to find an approximate analytical solution of (6), we expand the exponential $e^{-x^2/4Dt}$ into a power series:

$$e^{-\frac{x^2}{4Dt}} = 1 - \frac{x^2}{4Dt} + \frac{1}{2!}\left(\frac{x^2}{4Dt}\right)^2 + \dots + (-1)^n \frac{1}{n!}\left(\frac{x^2}{4Dt}\right)^n + \dots$$

(7)

The simplest equation for calculating the diffusion coefficient may be obtained by satisfying the condition

$$\frac{a^2}{4Dt} \ll 1.$$

(8)

In this case the exponential is close to unity, and Eq. (6) takes the form

$$I = \frac{I_0 \cdot a}{\sqrt{\pi Dt}} .$$

(9)

This result may also be obtained by supposing that the concentration of the diffusing material changes very little over the surface layer of thickness a. This is justifiable if the maximum depth of penetration of the diffusing element

$$x_{max} \gg a.$$

(10)

Since the layer completely absorbing the α radiation, a, is a small quantity, and for many metals equals 5 or 10 × 10^{-4} cm, conditions (8) and (10) can easily be satisfied experimentally. Then the computing formula for the diffusion coefficient obtained from expression (9) takes the form

$$D = \frac{a^2}{\pi t}\left(\frac{I_0}{I}\right)^2 .$$

(11)

Equation (11) contains an a^2, so that any error in this quantity may produce large errors in the value of the diffusion coefficient. This error clearly cannot be evaded by using tabulated values of a, since the effective width of the layer, the radiation of which is recorded by radiometric apparatus, depends on the presence of an air gap between the sample and the counter and on the sensitivity of the latter. Hence in calculating the diffusion coefficient from Eq. (11) we must either take exact account of all the factors affecting the absorption of α radiation or else determine the effective value of a experimentally.

The effective value of a can be determined if we know the integral activity of the sample before and after diffusion, I_0 and $I = f(x)$. In fact

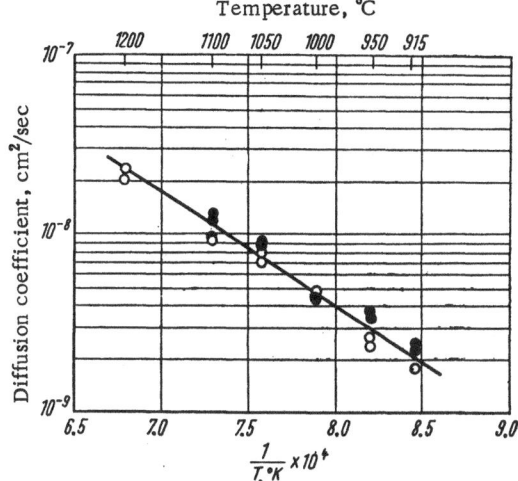

Temperature dependence of the diffusion coefficient of uranium in β titanium: $\bullet$) Absorption method; $\circ$) method of layer removal.

$$I_0 \sim c_0 h = \int_0^\infty c(x)\,dx = \sum_{n-1}^\infty \int_{a_n}^{a_{n+1}} c(x)\,dx \sim \sum_{n=1}^\infty \bar{I}_n.$$

The equation $I_0 = \sum_{n=1}^\infty \bar{I}_n$, where $\bar{I}_n$ is the average integral activity in the n-th layer of thickness a, is satisfied for a finite n, since in practice $I_n \to 0$ for $x \approx x_{max}$. In this case $n = x_{max}/a$. Thus, if we divide the diffusion layer into layers of equal thickness l, then for $l = a$ we have the relation

$$I_0 = \sum_{n=1}^{n=\frac{x_{max}}{a}} \bar{I}_n. \qquad (12)$$

By choosing the layer thickness in such a way that relation (12) is satisfied, we can find a.

In order to check the proposed method of determining the diffusion coefficient of α-active elements, we made an experimental study of the diffusion of uranium in β titanium. The diffusion coefficients were determined by the absorption method, using Eq. (11), and by the method of layer analysis, measuring the integral activity of the rest of the sample. For calculating the diffusion coefficient by the layer-analysis method we used Eq. (1). On satisfying the above conditions (regarding the negligible variation in the concentration over a layer of thickness a), we find that

$$\ln I = A - \frac{x^2}{4Dt}, \qquad (13)$$

where A is independent of x. The slope $\tan \alpha$ of the straight line $\log I = f(x^2)$ equals $\frac{1}{4}Dt$. Hence

$$D = \frac{1}{4t \cdot \tan \alpha}. \qquad (14)$$

The substance for study was 99.7% pure titanium remelted in an arc furnace in a pure argon atmosphere. From the bars of remelted titanium we prepared samples $8 \times 10 \times 16$ mm in size, annealing these at 1000°C for 5 h in vacuum in order to homogenize the structure. After additional preparation of the surface, uranium enriched with the U^{235} isotope was vacuum-deposited on the surface of the sample. The activity of the samples before and after diffusion annealing was measured with a B-2 radiometer furnished with a scintillation attachment. The accuracy of the radiometric system was checked by measuring the activity of a standard sample.

Diffusion annealing was carried out in a quartz capsule previously evacuated to 10^{-4} mm Hg. In order to prevent possible oxidation, the samples in the capsule were covered with titanium chips. The temperature was kept constant to ±5°C and measured with a platinum-platinorhodium thermocouple. In determining the diffusion coefficient by the layer-analysis method, parallel layers were removed on a grinding wheel. The thickness of the layer thus removed was measured with an optical gage to an accuracy of 2 or 3 μ.

The diffusion coefficients were determined at temperatures of 915, 950, 1000, 1050, 1100, and 1200°C. The duration of the diffusion-annealing periods was between 1 h at 1200 and 5 h at 915°C. During these periods the surface activity fell 15 to 20 times and the maximum depth of penetration of the uranium was 200 to 300 μ.

Table of Experimental Values of the Diffusion Coefficient
of Uranium in β-Titanium

No. of sample	Temperature, °C	Annealing time, h	$D. \times 10^{-9}$ cm^2/sec	
			absorption method using Eq. (11)	layer-analysis method using Eq. (14)
1 2	915	5	2.44 —	2.42 1.77
3 4	950	3	3.70 3.56	2.43 2.68
5 6	1000	1	4.35 4.57	— 4.77
7 8	1050	1	9.28 8.74	8.20 7.41
9 10	1100	1	12.2 13.0	9.20 9.86
11 12	1200	1	— —	20.4 22.8

The effective layer for the absorption of α radiation, a, found from relation (12) was 5.7×10^{-4} cm.

It may be shown that conditions (8) and (10) were adequately satisfied in the experiments, so that Eq. (11) was quite valid. Thus the ratio $a^2/4Dt$ calculated from the diffusion coefficients given below was 1 or 2×10^{-3}, and the ratio $a/x_{max} = 2$ or 3×10^{-2}.

The experimental values of the diffusion coefficients of uranium in titanium found by the absorption method and the method of layer analysis are given in the table and in the graph relating the diffusion coefficient to the reciprocal of the absolute temperature (see figure).

These results illustrate the fair agreement between the diffusion coefficients found by the absorption method and the method of layer analysis. We see from the figure that, within the limits of experimental error (15 to 20%), the points calculated by the two methods lie on a single straight line. The temperature dependence of the diffusion coefficient of uranium in β titanium may be represented by the equation

$$D = 5.1 \cdot 10^{-4} \exp\left(-\frac{29,300}{RT}\right). \tag{15}$$

Conclusions

1. A comparatively simple absorption method for determining the diffusion coefficient of α-radioactive elements has been described. Provided that $a^2/4Dt \ll 1$, the computing formula takes the form

$$D = \frac{a^2}{\pi t}\left(\frac{I_0}{I}\right)^2.$$

2. This method has been checked by studying the diffusion of uranium in β titanium, using both the absorption method and the method of layer-by-layer analysis. The experimental values of diffusion coefficient found by these two methods are in satisfactory agreement. The temperature dependence of the diffusion coefficient of uranium in β-titanium is represented by the equation

$$D = 5.1 \cdot 10^{-4} \exp\left(-\frac{29,300}{RT}\right) \text{ cm}^2/\text{sec.}$$

Literature Cited

1. Hevesy, G., and Seith, W. Z. Phys., 56:790 (1929).
2. Hevesy, G., and Seith, W. Z. Phys., 57:896 (1929).
3. Fürth, R. Handbuch der Physik und Techn. Mechanik, 7:667 (1930).
4. Hevesy, G., et al. Z. Phys., 79:197 (1932).
5. Seith, W., and Keil, A. Z. Metallk., 25:104 (1933).
6. Bochvar, A. A., et al. In book: Transactions of the Second International Conference on the Peaceful Use of Atomic Energy, Geneva, 1958. Contributions of Soviet scientists, Vol. 3, Moscow, Atomizdat, 1959, p. 370.
7. Adda, J., et al. J. Nucl. Mater., 120:(1956).
8. Adda, J., et al. J. Nucl. Mater., 300:(1959).

BEHAVIOR OF CARBON IN SYSTEMS OF THE
METAL—MOLTEN LITHIUM—CARBON TYPE

N. M. Beskorovainyi, V. K. Ivanov, and M. T. Zuev*

The corrosion resistances of construction materials in molten lithium is considerably affected by carbon impurities contained both in the lithium and the construction material itself. Iron, chromium, molybdenum, niobium, and other metals form carbides with carbon. Lithium also forms a carbide Li_2C_2 [1].

When construction metals come into contact with molten lithium containing carbon, or pure lithium touches carbon-containing materials, the behavior of the carbon will be determined by the relative chemical affinities of the metal and the lithium toward carbon.

A measure of the chemical carbon affinity of a metal is the free energy associated with the formation of the corresponding carbide (ΔF_T^0).

By comparing the free energies for the formation of the lithium and metal carbides, the probable behavior of the carbon in the $Me-Li(L)-C$ system may be determined.

Figure 1 shows the free energy for the formation of the carbides of a number of metals from the elements as a function of temperature; these are calculated from data presented in [2-4] and relate to 1 g-atom of carbon.

Special attention must be given to calculating the ΔF_T^0 for Li_2C_2, since the thermodynamic characteristics of lithium so far published include only its heat of formation from the elements $(\Delta H_{298} = -14.2 \text{ kcal/mole})$ [2]. By knowing the entropy of formation of lithium carbide, we could determine the value of ΔH_{298}^0 and to a first approximation also estimate the relationship $\Delta F_T^0 = f(T)$ [4].

The entropy of solids (calculated for 1 g-atom) may be determined [4] from the formula

$$S = \frac{3}{2} R \ln A_{av} + R \ln V_{av} - \frac{3}{2} R \ln T_m + a,$$

where A_{av} is the average atomic weight, i.e., the molecular weight divided by the number of atoms in the molecule, V_{av} is the average atomic volume, i.e., the average atomic weight di-

---

*Also taking part in the work were Engineers Yu. Ya. Tomashpol'skii, M. V. Teregulov, and A. F. Gekov.

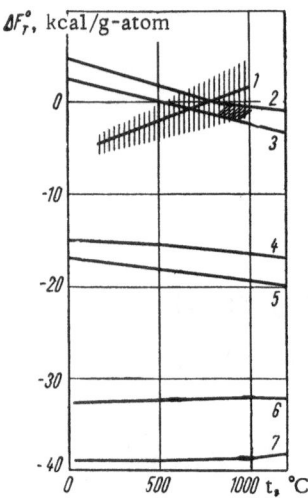

Fig. 1. Free energy of formation of the carbides of several metals as a function of temperature (shaded region corresponds to the error in calculating ΔF_T^0 for lithium carbide): 1) Li(L) + C(S) = $\frac{1}{2}$Li$_2$C$_2$(S); 2) 3Fe(S) + C(S) = Fe$_3$C(S) [2]; 3) 2Mo(S) + C(S) = Mo$_2$C(S) [2]; 4) $\frac{7}{3}$Cr(S) + C(S) = $\frac{1}{3}$Cr$_7$C$_3$(S); 5) (23/6)Cr(S) + C(S) = $\frac{1}{6}$Cr$_{23}$C$_6$(S) [2]; 6) Nb(S) + C(S) = NbC(S) [3]; 7) Ta(S) + C(S) = TaC(S).

vided by the density, T_m is the melting point, and a is a constant equal to 12.5 ± 2.

The melting point in this expression is not known for Li$_2$C$_2$. According to [1], Li$_2$C$_2$ dissociates vigorously at high temperatures; at 925°C the vapor tension of this compound is 0.35 kg/cm^2. Nevertheless, we may try to estimate the entropy of Li$_2$C$_2$ from Eastman's formula by taking the melting point of lithium carbide within reasonable limits, for example, 1300°K. This value was chosen by extrapolating the liquidus line on the phase diagram of the Li−C system [1]. The accuracy of determining the entropy from Eastman's formula is determined by the errors of the formula, ±2 cal/(g-atom · deg), and the error in selecting the melting point. On choosing the melting point of lithium carbide in the range 1300 ± 200°K, the error in determining the entropy does not exceed ±0.25 cal/(g-atom · deg). Thus the total error in determining the entropy of lithium carbide in our case is no greater than ±2.25 cal/(g-atom · deg) or ±9.0 cal/(mole · deg).

Using the value T_m = 1300°K and the values of the atomic weights and ρ = 1.65 g/cm^3 from [1], we obtain for lithium carbide S = 1.25 cal/(g-atom · deg) or S = 5.0 cal/(mole · deg).

Since the entropy is a positive quantity, the possible values of entropy must lie not in the range 5.0 ± 9.0 cal/(mole · deg) but in the narrower range $5.0^{+9.0}_{-5.0}$ cal/(mole · deg). We may well suppose that the range in which possible values of the entropy of lithium carbide lie is narrower than this. This conclusion may be reached from the following considerations.

According to the method of V. A Kireev [4], the entropy of a compound is expressed by the formula

$$S = \Delta S_a + \sum_1^k n_k (S_i)_k,$$

where ΔS_a is the entropy of formation of the compound from the elements when these are in the hypothetical state of a monatomic ideal gas, n_k is the number of atoms of a given element in the molecule of the compound, and $(S_i)_k$ are the entropies of the elements in the state of an ideal monatomic gas. Calculation of ΔS_a for Na$_2$C$_2$ from this formula gave −122 cal/mole·deg). Entropies of sodium and carbon equal to 12.3 and 1.36 kcal/(mole · deg), respectively [2], were used in the calculation, together with entropy of Na$_2$C$_2$ equal to −10.4 cal/(mole · deg) [5]; the entropy of sodium and carbon in the state of an ideal monatomic gas were calculated from data of [4].

For compounds of the same type, the values of ΔS_a within a single group of the periodic table rise with falling atomic number [4]. Thus the value of ΔS_a for Li$_2$C$_2$ should be greater in absolute magnitude than ΔS_a for Na$_2$C$_2$. Calculation by Kireev's method, using a lithium entropy of 6.75 cal/(mole · deg) [2] and the data of [4] indicates that in this case the entropy of Li$_2$C$_2$ should not exceed 9.7 cal/(mole · deg). Consequently, the possible values of the entropy of Li$_2$C$_2$ should not fall outside 5.0 ± 5.0 cal/(mole · deg).

From the known values of the entropies of lithium and carbon and the calculated value of the entropy of Li$_2$C$_2$, we can find the entropy of formation of Li$_2$C$_2$: $\Delta S = -11.2$

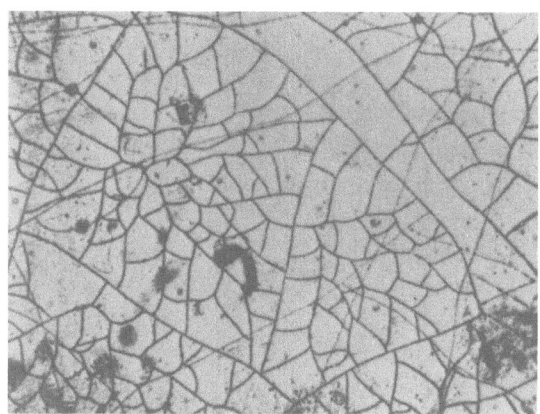

Fig. 2. Surface of molybdenum after tests at 1000°C in lithium for 220 h (× 45).

cal/(mole · deg). Using the value of the heat of formation of Li_2C_2 ($\Delta H_{298} = -14.2$ kcal/mole [2]) and the calculated value of the entropy of formation of Li_2C_2, we can obtain the $\Delta F_T^0 = f(T)$ relationship to a first approximation. Calculating for 1 g-atom of carbon, this is given by the formula

$$\Delta F_T^0 = -7100 + 5.6T \text{ cal/g-atom.}$$

This equation is valid for temperatures not exceeding the melting point of lithium. For higher temperatures we must allow for the change in the free energy of lithium on melting: $\Delta F = 690-1.5 T$ [2]. In this case the expression for the free energy of formation of lithium carbide, ΔF_T^0 takes the form

$$\Delta F_T^0 = -7800 + 7.1T \text{ cal/g-atom.}$$

For an error in the determination of the entropy of Li_2C_2 equal to ±2.5 cal/(g-atom · deg), the value of ΔF_T^0 will lie in the range −4.6 ± 1.1 kcal/g-atom or carbon at the melting point of lithium and in the range 1.2 ± 3.2 kcal/g-atom of carbon at 1000°C (shaded region in Fig. 1).

The value of ΔF_T^0 thus determined does not allow for the variation in the heat and entropy of formation of lithium carbide with temperature, and hence we cannot judge the possible behavior of the carbon in a system of the Me−Li(L)−C type unequivocally from these data. In addition to this, according to [2], the ΔF_T^0 curve suffers a downward break when phase transformations of carbides, oxides, nitrides, and other compounds occur. Since Li_2C_2 has three allotropic transformations [1], the curve of ΔF_T^0 for Li_2C_2 at high temperatures should lie rather below the calculated one, for example, as indicated in Fig. 1 by cross hatching in the 1000°C region.

The data shown in Fig. 1 regarding the ΔF_T^0 of carbide formation from the pure elements do not of course take into account the complex conditions of interaction to which construction materials are subjected during corrosion in molten-metal media such as lithium; nor do they account for the possible dissolution of the carbide phases in the solid and liquid metals, diffusion processes, and so forth. Considering that the accuracy of determining the ΔF_T^0 relationships for the majority of carbides lies at best between ±1 and ±3 kcal/mole, it would seem most appropriate to make a qualitative experimental estimate of the chemical affinity of metals toward carbon in a number of systems of the Me−Li(L)−C type.

We have therefore studied the effect of carbon impurities in lithium on iron, molybdenum, chromium, niobium, and tantalum. The experiments were all made under static isothermal conditions.

The Fe − Li(L) − C System

The mutual disposition of the free-energy curves corresponding to the formation of lithium and iron carbides (Fig. 1) indicate that at any rate at temperatures up to 800°C (the intersection point of the ΔF_T^0 curves for Fe_3C and Li_2C_2) lithium should have a higher chemical carbon affinity than iron. It was shown in [6] that at temperatures below 723°C lithium in fact decomposes Fe_3C with the formation of Li_2C_2.

It was indicated in the previous section that the ΔF_T^0 curve for Li_2C_2 (Fig. 1) should pass slightly below the computed curve; thus at temperatures above 800°C the thermodynamic stabil-

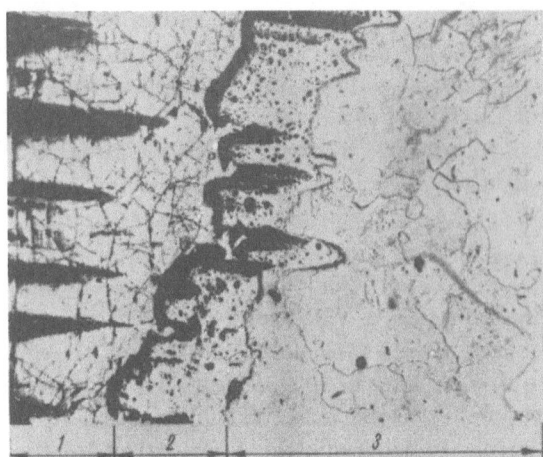
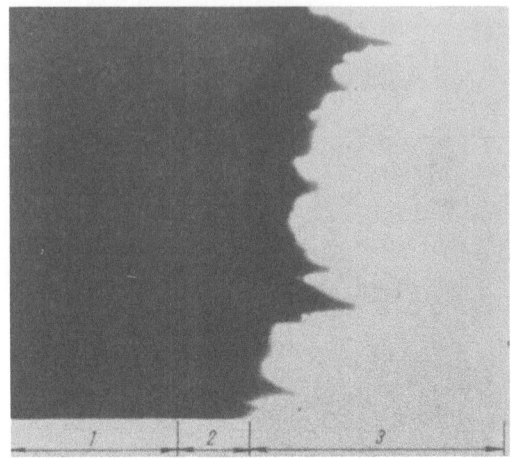

Fig. 3. Microstructure of an oblique section of molybdenum after tests in lithium at
1000°C for 220 h (×45): 1) Surface; 2) film;
3) metal.

Fig. 4. Autoradiogram of the oblique section of molybdenum shown in Fig. 3 (×45):
1) Surface; 2) film; 3) metal.

ities of lithium and iron carbides are apparently about the same, which makes a study of the
behavior of carbon in the $Fe-Li(L)-C$ system more difficult. The interaction of lithium with
$Fe-C$ alloys is considered in more detail in [6].

The Mo − Li(L) − C System

Molybdenum is well known as a fairly strong carbide-forming element, and its corrosion
resistance in lithium will thus be greatly determined by interaction with carbon contained in the
lithium. If the free energy for the formation of molybdenum carbides lies below the ΔF_T^0 for
Li_2C_2, then on contact between molten lithium and molybdenum we should expect that the carbon
impurity in the lithium would have an effect on the structure and properties of the molybdenum,
since in this case the formation of molybdenum carbides becomes possible. On the other hand,
if the free energy for the formation of molybdenum carbides lies above the ΔF_T^0 for Li_2C_2, then
the molybdenum may be decarburized, i.e., we may have an interaction analogous to that in the
$Fe-Li(L)-C$ system. In this case the carbon impurities in the lithium should not have any
serious effect on the structure and properties of the molybdenum.

Two molybdenum carbides Mo_2C and MoC, are known: the published thermodynamic
characteristics of these have a considerable spread. For example, in [2], values of ΔH_{298}^0 obtained from different sources are given; these are 15.8 and 4.2 cal/mole for Mo_2C and 9.7 and
2.0 kcal/mole for MoC. This makes an estimate of the ΔF_T^0 for the carbide formation of molybdenum more difficult.

According to [7], the carbide Mo_2C is formed when carbon diffuses into molybdenum. In
order to secure a qualitative idea of the behavior of carbon in the $Mo-Li(L)-C$ system, we
therefore used the ΔF_T^0 curve for Mo_2C. In order to plot this curve (Fig. 1) we choose a value
of $\Delta H_{298}^0 = 4.2$ kcal/mole as that found most frequently in the literature. The ΔF_T^0 curve for
Mo_2C lies rather below the ΔF_T^0 curve for Fe_3C and intersects the ΔF_T^0 curve for Li_2C_2 at approximately 700°C.

Hence at temperatures up to 700°C the Li_2C_2 should be the more stable, and a higher temperature this should apply to the Mo_2C.

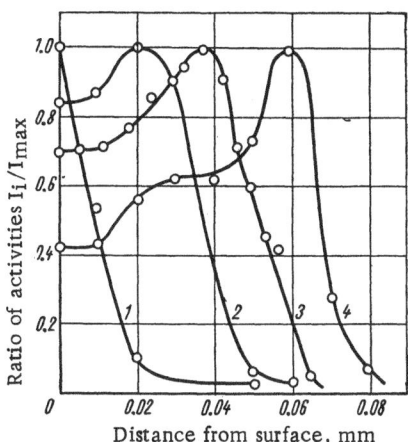

Fig. 5. Distribution of radioactive carbon in molybdenum after tests in lithium containing C^{14} at 1000°C for 1) 25; 2) 50; 3) 100; and 4) 220 h.

We only studied the interaction of molybdenum with the carbon impurities in lithium over a period of 25 to 220 h at 1000°C. The samples and the molybdenum containers were made of molybdenum rod containing some 0.03% carbon. The radioactive isotope C^{14} was introduced into the lithium in the form of iron-sensitive elements surfaced with C^{14}. After the tests the activity of the iron had fallen from 1 to 6×10^4 pulses · $(cm^2 \cdot min)^{-1}$ to 200 or 300 pulses/$(cm^2 \cdot min)$. Thus the carbon passed entirely into the lithium and according to calculation its concentration in the lithium reached approximately 0.2 wt.%.

After the tests the samples had increased in weight. The increment was around 475 mg/$(cm^2 \cdot year)$. The formation of a radioactive film on the molybdenum surface was established by radiometric layer analysis, autoradiography, and oblique microstructural analysis; the film had a microhardness $H_V = 700$ to 1100 kg/mm². The thickness of the film increased with duration of the test; after 220 h it was 0.07 mm. The film thus formed had a porous structure and its surface showed a network of grain boundaries; these may have been cracks (Fig. 2). Since the solubility of carbon in molybdenum is extremely low (10^{-5} to 10^{-4}% at 25°C and around 5×10^{-3} at 1500°C [8]), the film and the molybdenum had a sharp interface, as confirmed by the microstructure and autoradiogram of an oblique section (Figs. 3 and 4). The curves of carbon distribution in the film also had a sharp drop (Fig. 5). These results strongly suggest that a thin film of molybdenum carbides was formed on the surface of the molybdenum.

As indicated by the autoradiogram of the oblique section and the layer−by−layer radiometric analysis, the distribution of carbon in the carbide film takes place nonuniformly after 50 h of tests. The maximum concentration of carbon occurs at some distance from the sample surface. This distance changes parabolically with time according to the law

$$h \, [\text{mm}] = 4 \cdot 10^{-3} \sqrt{\tau} \, [\text{h}].$$

On the surface of the film, the carbon content falls as the test proceeds. This unusual distribution of carbon in the carbide film may be due to various causes.

First, plates of Armco iron were used during the tests in order to introduce that C^{14} tracer; this allowed iron atoms to be conveyed to the surface of the carbide film. This is supported by the fact that, on testing molybdenum carbide Mo_2C in lithium at 1100°C for 100 h in a reaction container made of Armco iron, the surface of the carbide showed a thin (5−μ) layer of a carbide of the Me_6C type [9]. The carbide Mo_2C is able to dissolve the iron carried from the sides of the iron container, and as the iron content increases the carbon content falls and the structure is altered. It is quite possible that in our case also the passage of iron to the surface of the carbide film produces a lower carbon content on the surface, together with a lower microhardness ($H_V = 700$ to 1000 kg/mm² as compared with 1500 kg/mm² for Mo_2C [10]).

The carbon distribution in the carbide film (in the form of a curve with a maximum) may also be associated with pore−forming processes [11]. In this case the maximum concentration of vacancies and micropores is usually found at some distance from the surface, and their distribution with respect to depth is described by a curve containing a maximum. The radioactive carbide film formed on the surface of the molybdenum after tests in lithium has a marked porous

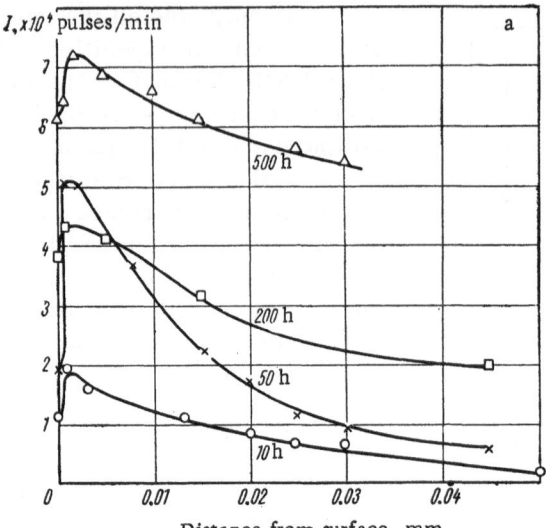

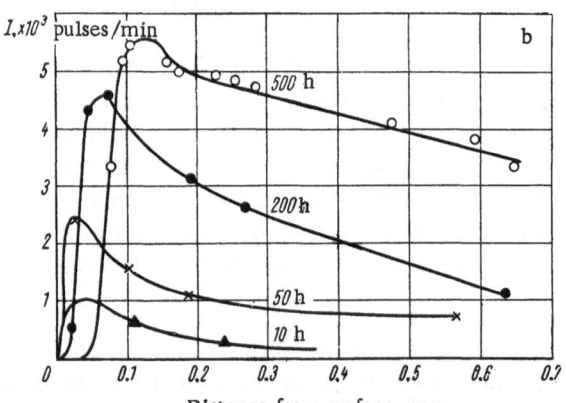

Fig. 6. Distribution of radioactive carbon in OKh12
steel after tests in lithium containing isotope C^{14} at
a) 600 and b) 800°C.

structure, and we may therefore suppose that the pores offer more favorable conditions for carbide formation (this also leads to a nonuniform carbon distribution).

It follows from all these data that at 1000°C molybdenum has a higher chemical affinity
for carbon than lithium, and hence the ΔF_T^0 curve for Mo_2C at 1000°C should lie below the
ΔF_T^0 curve for Li_2C_2 (see Fig. 1).

When molybdenum comes into contact with lithium containing carbon at 1000°C, carbide
formation should take place on the surface of the molybdenum.

The Cr − Li(L) − C System

It was shown in [12] that when chromium comes into contact with molten lithium containing carbon, there is a (so-called) transfer of carbon to the chromium. Samples of electrolytic
chromium tested in lithium containing the radioactive isotope C^{14} at 600 to 1000°C became radioactive to a depth of some 0.02 mm. The microhardness of the sample surfaces rose from 250
in the original state to 1000 or 1200 kg/mm². According to [10], the microhardness of chromium carbides $Cr_{23}C_6$ and CR_7C_6 equal 1650 and 1340 kg/mm²,respectively.

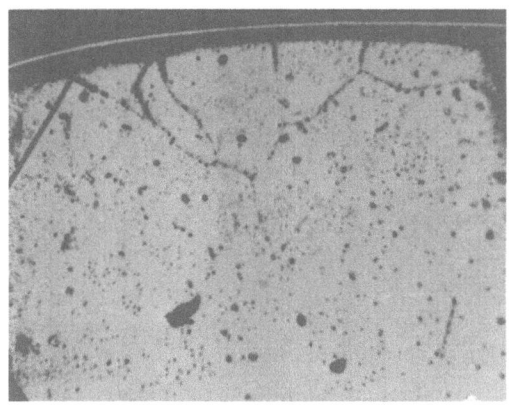

Fig. 7. Autoradiogram of a transverse section of OKh12 steel after tests in lithium containing radioactive isotope C^{14} at 600°C for 150 h ($\times$ 65).

These results suggest that the carbon in the lithium forms a thin film of carbides on the chromium surface. These carbides are clearly fairly stable in lithium, since on raising the temperature to 1000°C the radioactivity of the surface and the thickness of the activated layer increases.

The stability of the chromium carbides in molten lithium agrees with the data regarding the free energy of their formation presented in Fig. 1. On forming $Cr_{23}C_6$ and Cr_7C_3 in the temperature range 25 to 1100°C, ΔF_T^0 has values between -15 and -20 kcal/g-atom of carbon, while for Li_2C_2 ΔF_T^0 varies between -4.6 and $+1.2$ kcal/g-atom of carbon. Thus chromium has a higher chemical affinity toward carbon than does lithium, so that carbon passes from the molten lithium on to the surface of the chromium.

It is very important to estimate the chemical affinity for carbon of chromium in the case in which the chromium is in solid solution with iron, in particular for chromium stainless steels.

Tests on steels with low carbon contents (0.03 to 0.05%) and 12 to 17% chromium in molten lithium containing radioactive C^{14} show that, as in the tests with pure chromium, the samples become radioactive, and the activity of the lithium falls from 1 or 1.5 $\times$ 10^5 to 5 or 6 $\times$ 10^3 pulses/(cm^2 · min) at 600°C and to 4 or 7 $\times$ 10^2 pulses/(cm^2 · min) at 800°C. The distribution of radioactive carbon in the samples is shown in Fig. 6.

The fact that after tests in lithium containing C^{14} stainless chromium steels are activated and hence enriched with carbon, while the lithium is decarburized, indicates that even in solid solution with iron the chromium has a greater chemical affinity for carbon than does the lithium.

When carbon diffuses from lithium into chromium steel, the following processes may take place:

1) Dissolution of the carbon in the chromium α solid solution;

2) interaction of carbon with the chromium of the α solid solution, with the formation of chromium carbide;

3) interaction of the carbon with the chromium carbides present in the steel, with the formation of carbides containing more carbon or less chromium.

The first process may clearly only take place if the carbon content of the chromium α solid solution is below the solubility limit. If, however, the solubility limit of carbon in chromium ferrite is reached, then the diffusion of carbon from the lithium into the steel will lead to carbide formation.

The free energy of carbide formation from the elements (ΔF_T^0) changes, in the case of $Cr_{23}C_6$, from -17.7 to -18.0 kcal/g-atom of carbon as the temperature changes from 600 to 800°C; for Cr_7C_3 it changes from -15.7 to -16.1 kcal/g-atom of carbon [2]. The strength of the chromium α solid solutions, characterized by the free energy of mixing (ΔF_m) changes from -0.5 to -0.7 kcal/g-atom on changing the temperature from 700 to 800°C for a chromium content of 12 to 17% [13]. Allowing for this, the free energy of carbide formation from the ele-

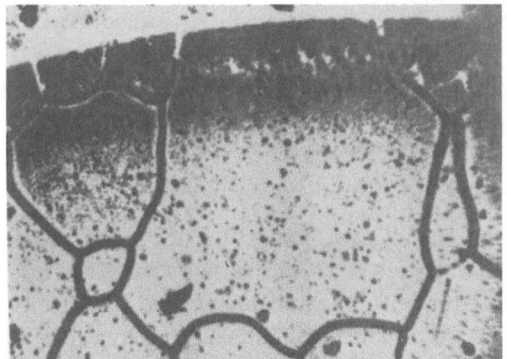

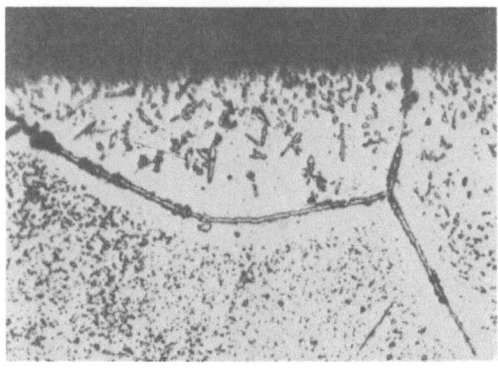

Fig. 8. Autoradiogram of a transverse section of OKh12 steel after tests in lithium containing radioactive isotope C^{14} at 800°C for 150 h ($\times$ 65).

Fig. 9. Microstructure of the transverse section of OKh12 steel shown in Fig. 8 ($\times$ 180).

ments when carbon interacts with the chromium of the α solid solution (ΔF) in the temperature range in question will vary between -15.7 and -15.2 kcal/g-atom of carbon for the reaction $^{23}/_6$ Cr + C = $^1/_6$ Cr$_{23}$C$_6$ ($\Delta F = \Delta F^0_T - ^{23}/_6 \Delta F_m$) and will be approximately -14.5 kcal/g-atom of carbon for the reaction $^7/_3$ Cr + C = $^1/_3$ Cr$_7$C$_3$ ($\Delta F = \Delta F^0_T - ^7/_3 \Delta F_m$).

The possibility of interaction between carbon and the carbides present in the steel, with the formation of carbides richer in carbon, at 600 to 800°C may be estimated for the reaction $^7/_{27}$Cr$_{23}$C$_6$ + C = $^{23}/_{27}$Cr$_7$C$_3$ from the values of ΔF, which are, respectively, equal to -12.5 and -13.1 kcal/g-atom of carbon; for the reaction $^3/_5$Cr$_7$C$_3$ + C = $^7/_5$Cr$_3$C$_2$ the value is approximately -3.4 kcal/g-atom carbon.

Comparison of the various values of ΔF given above for the rival processes indicates that when carbon diffuses into chromium steel at 600 to 800°C the most probable process is that of forming carbides Cr$_{23}$C$_6$ as a result of interaction between the carbon and the chromium of the α solid solution.

The carbide formation which, according to the thermodynamic calculations, should take place in chromium steel when carbon is present in the lithium, is confirmed by autoradiographic data, metallography, and microhardness measurements.

Thus, for example, at 600°C the radioactive carbon concentrates in a thin surface layer of the chromium steels (0.05 to 0.1 mm thick), and a certain amount of carbon penetrates along grain boundaries into deeper layers (Fig. 7). Since the diffusion processes are slow at 600°C, we find that, as indicated earlier, the lithium still contains a considerable amount of radioactive carbon after the tests (5000 to 6000 pulses/cm^2 · min), while the carbide phase appearing in the surface layers of the steel is highly dispersed. At 800°C, despite the fact that the carbon has passed almost entirely from the lithium into the steel, its concentration in the thin surface layer is zero, and only in the deeper layers is there an increased carbon concentration. In the zone with the increased carbon content, as in the case of the 600°C test, there is a carbide phase (Figs. 8 and 9), although the original structure consisted only of α phase grains, owing to the low carbon content (0.03 to 0.05%). We did not determine the structure and composition of the carbide phases formed; they may not have been purely chromium carbides, but carbides in which some of the chromium atoms were replaced by iron.

The carbide formation is confirmed not only by the microstructure but also by the increased microhardness of the zones in which the carbon concentration is increased (Fig. 10).

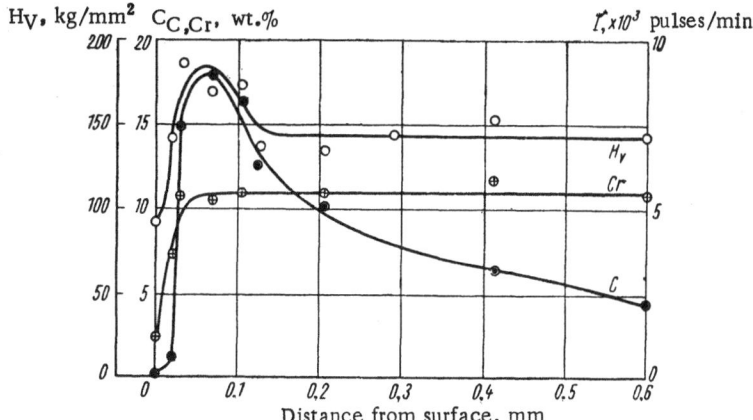

Fig. 10. Distribution of chromium and carbon in chromium steels after tests in lithium, together with the variation in microhardness with depth [9].

The absence of carbon from the thin surface layer is due to the solubility of chromium in lithium [9]. As the concentration of chromium in the surface layer diminishes, which is analogous to an increase in the iron content, the carbon should either pass into the lithium, for which the carbon affinity is higher than in the case of iron, or penetrate into deeper layers of the steel where there is still a high proportion of chromium. The latter process is thermodynamically more favorable, since, as indicated earlier, the chemical affinity of chromium for carbon is greater than that of lithium, even when the chromium is in solid solution with iron (see Figs. 6 and 10).

These data cast doubt on the explanation given for the results of tests carried out on 5Kh25 steel (in ampoules made of 1Kh13M2 steel) in lithium [9]. In these tests there was a fall in the amount of carbon in the 1Kh13M2 steel from 0.1 to 0.06% at a depth of 0.25 mm. It was considered that the carbon from the steel surface with the lower chromium content was carried to the 5Kh25 steel (which had the larger chromium content), and that some of the $Cr_{23}C_6$ carbides in the 5Kh25 sample were converted into Cr_7C_3. This explanation comes from the widely accepted view that carbon is transported through the lithium from low-chromium steels to those containing more chromium.

According to the data presented in the present paper, the following mechanism would appear more likely.

The first process is the dissolution in the lithium of a certain amount of chromium from the α solid solution of both the low- and high-chromium types of steel (1Kh13M2 and 5Kh25 for example). This follows from the fact that the chromium is bound much less strongly in the α solid solution than in the carbide phase; the strength of the carbides $Cr_{23}C_6$ and Cr_7C_3, calculated for one chromium atom, lies between −4.6 and −6.9 kcal/g-atom of carbon at 600 to 800°C, and that of the solid solution of chromium in iron is below −0.9 kcal/g-atom [13] at these temperatures.

When the chromium from the α solid solution dissolves in the lithium, the chromium equilibrium between the solid solution and the carbides is broken, which should lead to a redistribution of chromium from the carbide phase into the α solid solution, i.e., to dissolution of the carbides in the α solid solution. On dissolution, the crystal lattice of the chromium carbides breaks up, and, since the solubility of carbon in the α phase is insignificant, the carbon freed on decomposition of the carbides must pass down into the steel, into layers with high chromium contents, since chromium has a higher carbon affinity than lithium.

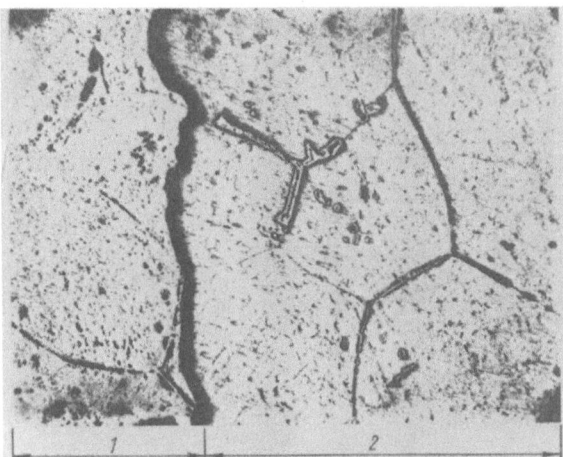

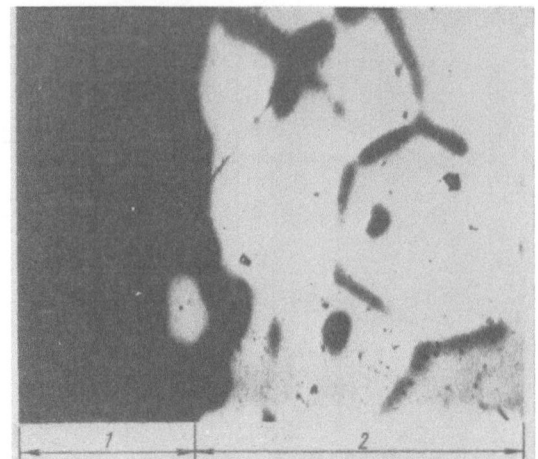

Fig. 11. Microstructure of an oblique section of niobium after testing in lithium at 1000°C for 113 h (×45): 1) Film; 2) metal.

Fig. 12. Autoradiogram of the oblique section of niobium shown in Fig. 11 (×45): 1) Film; 2) metal.

Thus we may assert that the carbon is not carried through the lithium from the low-chromium to the high-chromium steel, but passes into deeper layers of the same steel. This concept, however, requires further confirmation.

The carbide formation found on testing chromium and stainless chromium steels in molten lithium containing carbon indicates that chromium has a higher chemical affinity toward carbon than does lithium. This is true not only for pure chromium, but also for chromium in solid solution with iron, i.e., for the system $(Fe + Cr) - Li(L) - C$.

The $Nb - Li(L) - C$ System

Niobium forms carbides with carbon contents between 33.3 and 50 at.% [3]. On saturating metallic niobium with carbon, a film of carbide Nb_2C forms over the surface [7].

Judging from the data relating to the change in the free energy on forming carbides Nb_2C and NbC [3] (see Fig. 1), the thermodynamic strength of niobium carbides is 1.5 to 2 times higher than that of the chromium carbides. This is confirmed by data presented in [9], which show that on contact with molten lithium the carbides of niobium are characterized by high resistance.

An experimental study of the interaction between niobium and carbon in a molten-lithium medium was carried out with 99.6% pure niobium containing 0.12% carbon, Nb (I), and with 99.78% pure niobium containing about 0.01% carbon, Nb (II). The Nb (I) was tested for 25 to 250 h at 1000°C in containers of Armco iron. The radioactive isotope C^{14} was introduced into the lithium by means of iron plates. On contact with molten lithium at 1000°C, iron is well known to become decarburized; hence both the plates and the containers served as sources from which carbon impurity passed into the lithium.

During the tests, a radioactive film formed on the sample surfaces (Figs. 11-13); the microhardness of this increased with time and in 250 h reached 1460 kg/mm². According to [10], the microhardness of carbide Nb_2C is 2123 kg/mm² and that of carbide NbC, 1961 kg/mm². The film has sharp boundaries, which agrees with information regarding the insignificant solubility of carbon in niobium (some 0.015 wt.% C at 1000°C [8]).

The carbon concentration remains constant over the whole thickness of the film, indicating that the film is of constant composition (see Fig. 13).

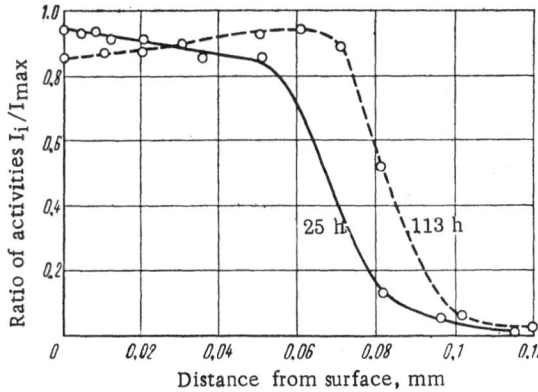

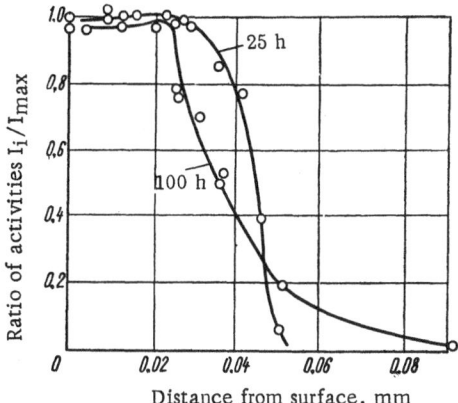

Fig. 13. Distribution of radioactive carbon in niobium after tests in lithium containing radioactive isotope C^{14} at 1000°C.

Fig. 14. Distribution of radioactive carbon in tantalum after tests in lithium containing radioactive isotope C^{14} at 1000°C.

From all these results we may conclude that the carbon in the lithium forms a carbide with niobium on the surface in contact with the lithium. The process of carbide formation develops along grain boundaries also, as may be seen from the microstructure (Fig. 11) and autoradiogram (Fig. 12). This arrangement of carbides in unfavorable to the mechanical strength of niobium.

Samples of Nb (II) were studied at 1100°C in lithium for a period of 100 h, the lithium being activated by the direct introduction of radioactive isotope C^{14}. The containers for these tests were made of Armco iron, molybdenum, and niobium.

After testing the Nb (II) in iron containers, the carbon penetrated into the surface layers of the sample [as in the case of Nb (I)]; this was confirmed by layer-by-layer radiometric analysis. The thickness of the radioactive layer for test periods of 100 h was 0.05 to 0.06 mm, and the microhardness was 2250 kg/mm². X-ray structural analysis of the film showed that it had the structure of a bc cube with lattice parameters a = 4.40 A, which may correspond to carbide NbC (H_V =1960 kg/mm², a = 4.469 A [10]).

After testing Nb (II) in molybdenum containers, a thicker radioactive film developed: 0.01 mm thick, with a microhardness of 1130 kg/mm².

On testing in niobium containers, despite the presence of radioactive radiation on the sample surface, no film was observed. The microhardness of the surface was H_V = 150 kg/mm², whereas the original microhardness of the samples was H_V = 160 kg/mm².

The results of the tests showed that carbide formation in the Nb (II) samples depended on the amount of carbon in the lithium. On testing in iron containers, all the carbon from the lithium and also that from the Armco iron passed into the niobium. On using molybdenum containers, two carbide-forming agents, each with a higher carbon affinity than lithium, entered the system (see Fig. 1), and the carbon present in the lithium distributed itself between the molybdenum and niobium in accordance with their relative affinities. Hence in this case the surface of the niobium sample contained less carbon than on testing in iron containers. On using a niobium container the area of the surface absorbing carbon increased greatly, and hence the carbon from the lithium, together with isotope C^{14}, was distributed in a very thin layer, and despite the high radioactivity of the surface no continuous film appeared. The absence of a continuous carbide film meant that the microhardness of the samples tested in niobium containers was almost unaltered.

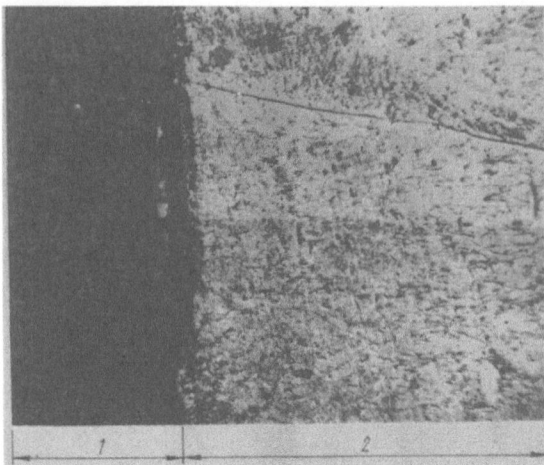

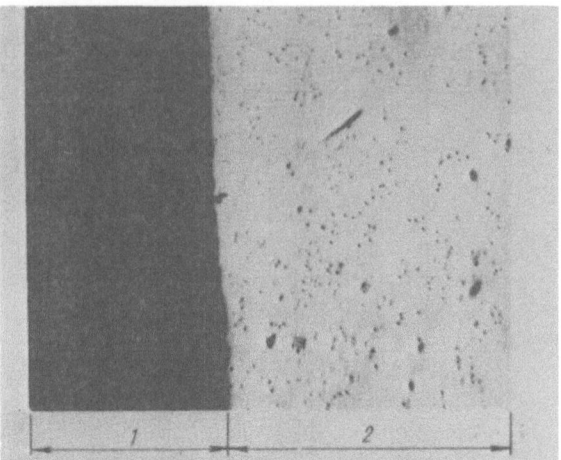

Fig. 15. Microstructure of an oblique section of tantalum after tests in lithium at 1000°C for 100 h (×45): 1) Film; 2) metal.

Fig. 16. Autoradiogram of the oblique section of tantalum shown in Fig. 15 (×50): 1) Film; 2) metal.

This situation is supported by the weight increments in the niobium samples after testing. On testing in iron containers the increment was 50 mg/cm^2, in molybdenum containers 2 mg/cm^2, and in niobium containers quite negligible.

The results thus obtained for the Nb−Li(L)−C system lead to the following conclusions.

Carbon impurities in lithium at 1000 and 1100°C cause carburization of niobium surfaces. If there is a considerable amount of carbon in the lithium, this may lead to the formation of a continuous carbide film on the sample surface, probably consisting of carbide NbC, and to the formation of niobium carbides along the grain boundaries. The behavior of carbon in the Nb−Li(L)−C system agrees with the mutual disposition of the curves in Fig. 1 describing the temperature dependence of the free energies of formation of niobium and lithium carbides.

The Ta − Li(L) − C System

Like the niobium carbides, the tantalum carbides Ta$_2$C and TaC have high thermodynamic strength (see Fig. 1).

The interaction of tantalum with carbon in molten lithium was studied on samples prepared from 99% pure sheet-tantalum material. The samples were tested in Armco-iron containers at 1000°C for 25 and 100 h. The lithium was activated by the introduction of an iron plate previously surfaced with isotope C^{14}.

Layer-by-layer radiometric analysis showed that the carbon in the lithium penetrated the surface layers of the tantalum (Fig. 14). The sample surfaces thereupon develop a golden-colored film, which is characteristic of tantalum carbide TaC, although according to [7] Ta$_2$C forms on metallic tantalum when this is saturated with carbon. The sharp boundary of the carbon-saturated zone in the microstructure of Fig. 15 and the autoradiogram (Fig. 16) indicate that there is a very restricted solubility of carbon in tantalum (0.01% at 1000°C [8]). The microhardness of the film after 100-h testing equals 1360 kg/mm^2, which is close to the microhardness of TaC (1600 kg/mm^2) [10].

The results indicate that tantalum has a higher chemical affinity for carbon than does lithium, as indeed we should expect on the basis of the mutual dispositions of the free-energy curves relating to the formation of the corresponding carbides (Fig. 1).

Conclusions

The results which we have been describing constitute an attempt to estimate the behavior of carbon when various constructional metals are placed in contact with molten lithium, using the thermodynamic stabilities of the corresponding carbides represented by the change in free energy taking place when the carbides are formed. The ΔF^0_T curves calculated for reactions in the solid phase without considering the solubility of carbon in the metals give some idea of the processes taking place in the Me−Li(L)−C system, at any rate in those cases in which the thermodynamic stability of the metal carbides differs considerably from that of lithium carbide. The use of radioactive isotope C^{14} provides a qualitative estimate of the relative disposition of the ΔF^0_T curves corresponding to carbide formation.

Literature Cited

1. Grishin, V. K., et al. Properties of Lithium, Moscow, Metallurgizdat, 1963.
2. Krestovnikov, A. N., et al. Handbook on Calculating the Equilibrium of Metallurgical Reactions, Moscow, Metallurgizdat, 1963.
3. Gerasimov, Ya. I., et al. Chemical Thermodynamics in Nonferrous Metallurgy, Vol. 3, Moscow, Metallurgizdat, 1963.
4. Karapet'yants, M. Kh. Chemical Thermodynamics, Moscow-Leningrad, Goskhimizdat, 1953.
5. Reports of the United States Atomic-Energy Commission. Nuclear Reactors. Vol. 3. Materials for Nuclear Reactors [Russian translation], IL, 1956.
6. Beskorovainyi, N. M., and Ivanov, V. K. Mechanism underlying the corrosion of carbon steels in lithium (this volume, p. 121).
7. Samsonov, G. V. Fiz. Metal. i Metalloved., 2:309 (1956).
8. Hahn, G. T., et al. The effects of solutes on the ductile-to-brittle transition in refractory metals. In: Refractory Metals and Alloys, Vol. 17 of Metallurgical Society Conferences, Interscience, New York, 1963, p. 24.
9. Nevzorov, B. A., et al. Study of the Corrosion Resistance of Constructional Materials in Alkali Metals. Paper No. 343 (USSR) presented to the Third International Conference on the Peaceful Use of Atomic Energy, Geneva (1964).
10. Samsonov, G. V. Refractory Compounds, Moscow, Metallurgizdat, 1963. [English edition: Handbook of High-Temperature Materials, Plenum Press, New York, 1964.]
11. Geguzin, Ya. E. Macroscopic Defects in Metals, Moscow, Metallurgizdat, 1962.
12. Beskorovainyi, N. M., and Yakovlev, E. I. In collection: Metallurgy and Metallography of Pure Metals, No. 2, Moscow, Atomizdat, 1960, p. 189.
13. Lesnik, A. G. Vopr. Fiz. Metal. i Metalloved., Akad. Nauk Ukr. SSR, Sb. Nauchn. Rabot, 1960, p. 18.

MECHANISM UNDERLYING THE CORROSION
OF CARBON STEELS IN LITHIUM

N. M. Beskorovainyi and V. K. Ivanov*

It was shown in [1, 2] that carbon steels become decarburized in molten lithium and that as a result of this their structure and properties undergo severe changes. An explanation of the processes taking place when carbon steels come into contact with molten lithium may prove useful when studying the corrosion of other carbon-containing materials in a molten-metal medium.

It was shown earlier [1] (by introducing radioactive C^{14} into lithium (containing up to 25 at.% carbon in all) that there was no diffusion of carbon into the iron, either in the α or γ region, nor was there any formation of iron carbides or dissolution of carbon in γ iron. On the other hand, when carbon steels containing 0.1 to 1.2% carbon were tested in lithium at 600 to 1000°C with isothermal holding periods up to 100 h, the carbide phase Fe_3C and the γ solid solution always tended to break up. Hence lithium has a greater chemical affinity for carbon over a wide temperature range than does iron.

Thermodynamic calculations [3] show that, at any rate up to 723°C, when carbon is present in steel in the form of cementite (Fe_3C), the lithium may react with the cementite in accordance with the reaction:

$$2Li(L) + 2Fe_3C(S) = Li_2C_2(S) + 6Fe(S). \tag{1}$$

The possibility of reaction (1) taking place at the temperatures corresponding to the existence of cementite is confirmed by the following data, obtained by studying the diffusion of lithium into the surface layers of carbon steels:

1) The depth of penetration of the lithium into the steel coincides with the depth at which decomposition of the pearlite occurs;

2) the curves giving the distribution of lithium in carbon steels (Fig. 1a and b) have a sharply bounded region with a constant lithium concentration (horizontal part), typical of the formation of a chemical compound by reactive diffusion; the lithium concentration in this region corresponds to the stoichiometric concentration of lithium in Li_2C_2 (Table 1).

These facts indicate that reaction (1) does in fact take place at 700°C, when the carbon is present in the steel as Fe_3C. Analogous analysis of data presented in [2] leads to the same conclusion for 600°C.

The lithium carbide formed by reaction (1) should not remain unchanged on contact with the molten lithium surrounding the sample. This follows from the phase diagram of the Li—C

*Engineer Chang Chia-shou took part in the work.

TABLE 1. Concentration of Lithium in the Corrosion Zone of Carbon Steels with Various Compositions after Tests in Lithium at 700°C

Type of steel	Carbon content of steel, wt.%	Computed lithium content for steel of the composition in question, corresponding to the stoichiometry of Li_2C_2, wt. %	Experimental lithium concentration, wt. %
St.20	0.21	0.12	0.11–0.12
St.45	0.46	0.265	0.25–0.27
U12	1.22	0.71	0.72–0.74

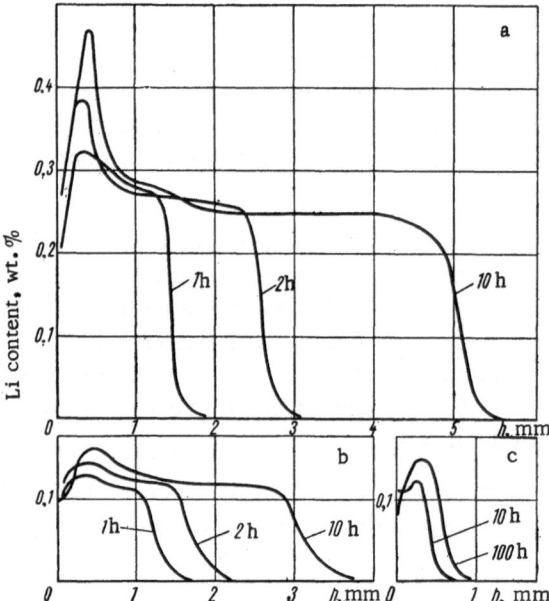

Fig. 1. Distribution of lithium with respect to depth in various types of steel: a) St.45; b) St.20 (tested in lithium at 700°C); c) St.20 tested at 900°C.

system (Fig. 2). According to this diagram, lithium carbide dissolves in molten lithium. On increasing the temperature, the amount of the solid phase, Li_2C_2, falls continuously, and the amount of the liquid phase (solution of carbon in lithium) and its carbon content corresponding increase. Thus the dissolution of Li_2C_2 in lithium should lead to the appearance of a liquid phase (solution of carbon in lithium) in the corrosion zone; this phase contains, for example, up to 5 wt.% carbon at 700°C. The carbon concentration gradient between this phase and the molten lithium surrounding the sample should lead to a redistribution of carbon, the carbon content of the corrosion zone diminishing and the molten lithium becoming richer in carbon.

Experimental data show that the penetration of lithium into the corrosion zone and the formation of Li_2C_2 are in fact accompanied by the decarburization of this zone, as indicated by the autoradiograms of steel treated with radioactive C^{14} after tests in lithium at 600°C

TABLE 2. Calculation of the Volumes of the Phases Corresponding to Reactions (1) and (2) under Standard Conditions

Reaction	Reagent	Wt., g	Density, g/cm³	Volume of phases, cm³
$2Li + 2Fe_3C = Li_2C_2 + 6Fe$	$2Li$	13.9	0.53	26.2
	$2Fe_3C$	359.1	7.67	46.8
	Li_2C_2	37.9	1.65	23.0
	$6Fe$	335.1	7.87	42.6
$Li_2C_2 \rightarrow 2Li + 2C$	Li_2C_2	37.9	1.65	23.0
	$2Li$	13.9	0.53	26.2
	$2C$	24.0	2.2	10.9

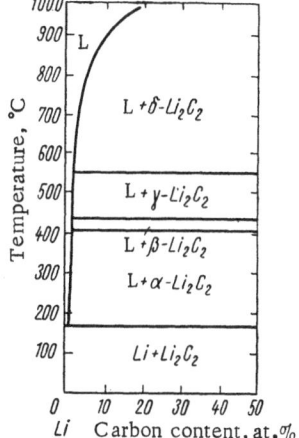

Fig. 2. Phase diagram of the Li−C system [5].

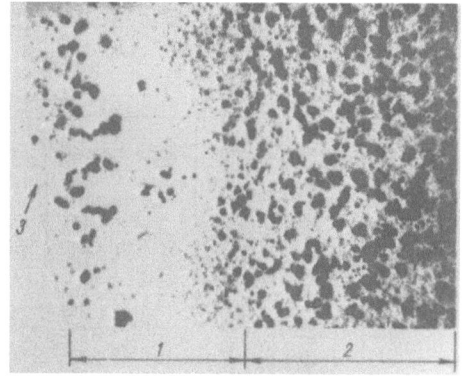

Fig. 3. Autoradiogram of a transverse section of U7 steel treated with radioactive C^{14}, after testing in lithium at 600°C for 2 h (×65): 1) Corrosion zone; 2) noncorroded material; 3) surface.

(Fig. 3). As time progresses, the depth of the decarburized zone increases; hence the dissolution of Li_2C_2 and the transfer of carbon to the lithium surrounding the sample begins from the surface and gradually propagates into deeper layers of the steel.

Metallographic study of the corrosion zone of carbon steels shows that the products of the reactions taking place in the corrosion zone are disposed along the grain boundaries of the ferrite in the form of dark inclusions and microchannels. Since the formation and subsequent dissolution of the Li_2C_2 leads to the appearance of a liquid phase in the corrosion zone, the microchannels containing the liquid phase may serve as routes along which decarburization of the corrosion zone takes place, together with the diffusion of lithium into deeper layers of the steel.

It is known that the volume of the carbon steel increases after tests in lithium, while the density falls [2]. It was suggested in [2] that the fall in density was related to the formation of corrosion products in the surface layers of the steel, these products having a greater specific volume than the original material.

The calculation summarized in Table 2 shows that, when lithium diffuses into steel and reacts with the cementite in accordance with reaction (1), the volume of the final products

TABLE 3. Volume Changes in Carbon Steels Tested
in Lithium at 700°C

Type of steel	Length of test, h	ΔV_{calc}, mm³	ΔV_{exp}, mm³
St. 20	1 2 10 100	75 105 160 225	75 100 150 250
St. 40	1 2 10 100	135 185 450 500	130 150 405 620
U 12	1 2 10 100	355 430 1240 1240	310 360 1050 1240

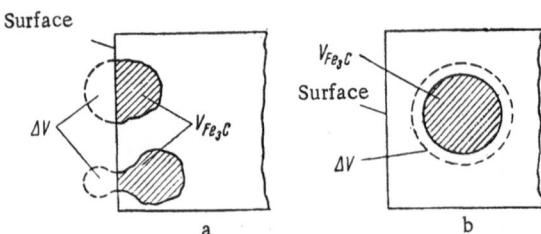

Fig. 4. Volume changes accompanying reactions
(1) and (2): a) At the surface; and b) in the depths
of the corrosion zone.

(65.6 cm³) is roughly 40% greater than that of the original cementite (46.8 cm³). In addition
to this, there should also be volume changes when the Li_2C_2 formed dissolves in the molten
lithium. The value of these may to a first approximation be estimated from the change in volume taking place when Li_2C_2 dissociates, the point being that on dissolution in lithium the Li_2C_2
breaks up thus:

$$Li_2C_2 \text{ (S)} \rightarrow 2Li \text{ (L)} + 2C \text{ (S)}. \tag{2}$$

We see from Table 2 that the volume of the dissociation products exceeds that of the
Li_2C_2 molecules by 14.1 cm³, corresponding to some 30% of the original cementite volume.

Thus the over-all volume changes associated with reactions (1) and (2) equal roughly
70% of the volume of the cementite in the corrosion zone.

Because iron has a low yield point at 600 to 700°C, these volume changes may produce
plastic deformation of the corrosion zone and an increase in sample volume. Knowing the volume of the corrosion zone and the amount of carbon in the steel, we can calculate by how much
the sample volume should increase, if we assume that reactions (1) and (2) and the corresponding volume changes in the corrosion zone pass to completion:

$$\Delta V_{calc} = 1.37 \cdot V \cdot \rho \cdot C,$$

where V is the volume of the corrosion zone, cm³, ρ is the density of the steel, g/cm³, C is
the carbon content of the steel, wt.%, and 1.37 is a coefficient equal to the change in volume

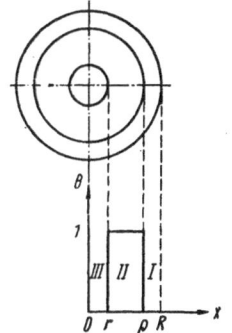

Fig. 5. Distribution of volume changes (θ) over the depth of the corrosion zone. In region I there are no volume changes (see Fig. 4a); in region II the changes are incomplete (see Fig. 4b); region III corresponds to the uncorroded part of the sample.

(cm^3) corresponding to 1 g of carbon when reactions (1) and (2) pass to completion (according to Table 2, 24 g of carbon taking part in the reaction produce a volume change of $18.8 + 14.1 = 32.9$ cm^3).

A calculation of this kind was carried out for a number of carbon steels tested at 700°C. The results, presented in Table 3, show that the calculated volume changes (ΔV_{calc}) have approximately the same value as those found experimentally (ΔV_{exp}). The agreement between these two values indicates that corrosion does in fact take place by the mechanism proposed. The slightly smaller experimental value is probably due to the fact that the theoretical value was computed on the assumption that reactions (1) and (2) were completely carried out in the corrosion zone; this is probably only true for fairly long tests (100 h). It was also assumed in the calculation that the volume changes corresponding to reactions (1) and (2) took place completely in the corrosion zone.

If, however, inclusions of the original phase, for example, Fe_3C, are disposed directly at the surface of the sample, then the reaction products for which there is no room in the original volume have nothing to stop them passing outside the sample (Fig. 4a). Hence reactions (1) and (2) may take place, in effect, without volume change in the surface layer of the corrosion zone, and only in deeper layers of the corrosion zone, from which the excess corrosion products cannot pass out of the sample, will reactions (1) and (2) produce volume changes (Fig. 4b). Thus, the experimental value of the volume changes may fall below that determined theoretically for two reasons:

1) the lagging of reaction (2) on reaction (1), as a result of which the lithium carbide formed is for some while kept partly in the corrosion zone;

2) the different conditions governing reactions (1) and (2) at the surface and well inside the corrosion zone, as a result of which the volume changes corresponding to reactions (1) and (2) only occur in layers some distance from the steel surface, and are absent at the surface itself.

Because reactions (1) and (2) produce no volume changes at the sample surface, the lithium concentration should be reduced in this region. This is shown by the following calculations. The original volume occupied by the cementite equals 46.8 cm^3 (see Table 2). After reactions (1) and (2) have taken place, part of this volume (42.6 cm^3) will be occupied by iron. Since reactions (1) and (2) take place without volume change in the surface layer, the maximum volume which can be occupied by lithium is $46.8 - 42.6 = 4.2$ cm^3. In this case the lithium content at the surface equals 4.2 times 0.53, or 2.2 g, i.e., some 16% of the amount of lithium corresponding to the formation of Li_2C_2 by reaction (1).

On the other hand, in deeper layers of the corrosion zone, where volume changes take place exactly in accordance with reactions (1) and (2), the lithium content should be above that corresponding to the formation of Li_2C_2 by reaction (1). The volume which the products of reactions (1) and (2) actually occupy equals $42.6 + 26.2 + 10.9 = 79.7$ cm^3 (see Table 2). If we suppose that, when the carbon passes from the corrosion zone into the molten lithium surrounding the sample, the volume thus freed (10.9 cm^3) is able to fill up with lithium coming into the sample in excess of the quantity required to form Li_2C_2, then the lithium content in the completely decarburized layers will rise by 10.9 times 0.53, or 5.8 g, i.e., some 42% of the amount of lithium corresponding to the formation of Li_2C_2 by reaction (1).

TABLE 4. Additional Quantity of Lithium Entering the Steel
on Decarburization of the Corrosion Zone at 700°C

Type of steel	Length of test, h	P_{tot}, mg	$P_{Li_2C_2}$, mg	ΔP_{exp}, mg	ΔP_{calc}, mg
St. 20	1	30	30	0	10
	2	40	35	5	15
	10	65	60	5	25
	100	125	95	30	40
St. 45	1	50	50	0	20
	2	75	65	10	30
	10	190	180	10	75
	100	290	210	80	85
U12	1	170	150	20	65
	2	250	185	65	75
	10	530	415	115	170
	100	620	445	175	180

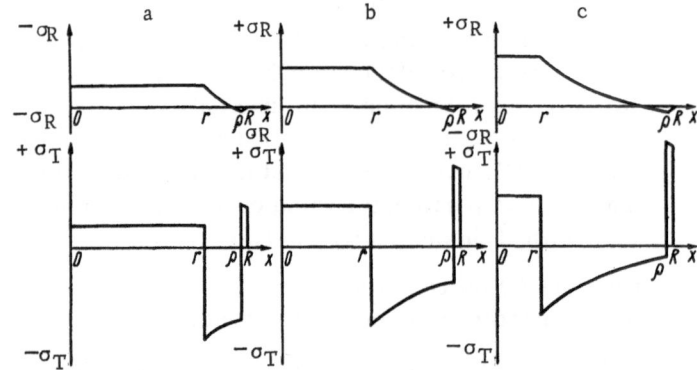

Fig. 6. Distribution of elastic stresses over the depth
of the corrosion layer (cf., Fig. 5) for various moments
of time (t) (σ_R = radial stresses, σ_T = tangential stresses):
a) $\rho = 0.96\ R$, $r = 0.75\ R$, t_1; b) $\rho = 0.96\ R$, $r = 0.5\ R$,
t_2 ($t_2 > t_1$); c) $\rho = 0.96\ R$, $r = 0.25\ R$, t_3 ($t_3 > t_2 > t_1$).

In accordance with the above discussion, the form of the lithium-distribution curves over
the depth of the corrosion zone in carbon steels should differ from the classical forms of curves
characterizing reactive diffusion. In the surface layers the lithium concentration should be reduced, and in the layers where decarburization has taken place it should be increased, i.e., the
curve should show a maximum. The experimental curves shown in Fig. 1a and b have precisely
this form. The curves contain a maximum, the size of which increases as decarburization of
the corrosion zone proceeds. The layer-analysis method which we used to determine the lithium content [2] enabled us to analyze a layer 0.1 mm thick but not to establish the lithium concentration directly at the sample surface; the general run of the experimental curves nevertheless indicated that the lithium content at the surface was below that corresponding to the
formation of Li_2C_2. Thus the nature of the lithium distruction over the depth of the corrosion
zone agrees with the proposed mechanism governing the penetration of lithium into steel.

Calculation shows that the total amount of lithium in the corrosion zone, obtained by integrating the diffusion curves over the sample volume, exceeds the amount necessary for the
formation of Li_2C_2 by a quantity

Fig. 7. Failure of U12 steel samples after 2-h tests at 700°C in a) pure lithium and b) lithium containing 10 at.% carbon.

$$\Delta P_{exp} = P_{tot} - P_{Li_2C_2},$$

where P_{tot} is the lithium content of the sample obtained by integration of the diffusion curves in mg, $P_{Li_2C_2}$ is the calculated amount of lithium required for the formation of Li_2C_2 in the corrosion zone.

This fact confirms the view that, when the corrosion zone is decarburized, an extra quantity of lithium can pass into it, the actual amount depending on the degree of decarburization. It was pointed out earlier that complete decarburization of the corrosion zone only occurred for long tests; we should therefore expect that for long tests ΔP_{exp} would coincide with the values of ΔP_{calc} obtained on the assumption of complete decarburization of the corrosion zone. For shorter tests the experimental values should fall below the theoretical.

Table 4 shows values of ΔP_{exp} for a number of carbon steels and the corresponding values of ΔP_{calc}, equal to the amount of lithium which would occur in the corrosion zone if reactions (1) and (2) passed to completion in this region.

The data of Table 4 show the amount of extra lithium coming into the carbon steels during fairly short tests in fact corresponds to incomplete decarburization of the corrosion zone, while for 100-h tests the corrosion zone is almost entirely decarburized. These results, together with the data presented in Table 3, indicate that there is a factor in the corrosion zone which retards decarburization, or, in other words, impedes the flow of reaction (2).

Such a factor might be a complex stresses state arising in the carbon-steel samples on corrosion in lithium as a result of the volume changes taking place. The distribution of stresses over the depth of the corrosion zone may be determined if we know the solution of the elastic-plastic problem. Since no such solution has been published for our case, we may attempt an approximate determination of the stressed state of the sample by solving the elastic problem. The solution of this problem for the case presented in Fig. 5 shows that in the surface layer of the corrosion zone there are tensile tangential and radial stresses, while at deeper levels there are tensile radial and compressive tangential stresses, ten times exceeding the yield point of iron at 600 to 700°C (Fig. 6). Hence the corrosion zone should be deformed plastically under the influence of these stresses, as we observed. The surface layer should be in tension and the deeper layers in compression. The pressure exerted by the compressed layers on the Li_2C_2 inclusions may cause the retardation of reaction (2) and hence the lagging of the decarburization of the corrosion layer relative to its growth.

The existence of considerable tensile stresses in the surface layers may produce the cracking sometimes found in the surfaces of the carbon-steel samples tested (Fig. 7).

The intensity of these corrosion processes taking place in carbon steels tested in molten lithium should depend on the amount of carbon in the lithium as well as that in the steel. As the amount of carbon in the lithium increases, we should expect, on the basis of the corrosion mechanism proposed, that the decarburization of the steel will take place less intensely in view of the small carbon concentration gradient, and that the volume changes in the dissolution of Li_2C_2 and hence the access of extra lithium to the corrosion zone will be smaller, since these depend on the degree of decarburization.

The results of tests carried out on carbon steels in lithium containing 10 to 30 at.% carbon at 700°C show that smaller changes take place in the volume of the steel in this case, and there are therefore less sharp changes in the strength and ductility (Table 5).

TABLE 5. Effect of the Carbon Content in the Lithium on the
Volume Changes in the Carbon Steel and on Its Mechanical
Properties

Type of steel	Medium	Length of test, h	ΔV_{exp}, mm³	σ_B, kg/mm²	δ, %
St. 20	Before testing	—	—	51.7	37.4
	Li	1 10 100	75 150 250	16.9 10.7 8.8	19.3 3.6 3.0
	Li + 10 at. % C	100	240	9.3	4.2
	Li + 30 at. % C	1 10 100	65 145 210	24.2 18.9 18.1	24.1 18.9 5.9
St. 45	Before testing	—	—	75.1	38.2
	Li	10	405	15.2	3.6
	Li + 30 at. % C	10	330	22.3	4.5

Thus the experimental data confirm the proposed mechanism for the rupture of carbon steels in lithium at temperatures up to 723°C, for which the carbon in the steel takes the form of Fe_3C.

At higher temperatures (for example, 900 to 1000°C), the carbon is in the γ solid solution; hence any processes taking place in carbon steels tested in lithium will differ from those corresponding to lower temperatures. Since there is practically no diffusion of lithium into pure iron [2, 4], interaction between lithium and the carbon in the γ solid solution at high temperatures can only take place at the surface of the steel, i.e., where the lithium and carbon are in direct contact.

Hence the intensity of the interaction should be determined by the rate at which carbon atoms come to the surface from inside the sample.

Experimental data confirm the view that lithium meets with difficulty in diffusing into carbon steel at 900 to 1000°C; its depth of penetration is greatly reduced (Fig. 1c).

The lithium concentration in the surface layer at high temperatures is comparable in magnitude with that of lithium in carbon steels tested at lower temperatures. This may confirm the identity of the processes taking place in the steel at 700 and 900°C. According to the Li −C phase diagram, however, the solubility of carbon in lithium rises considerably at high temperatures, while the thermodynamic stability of Li_2C_2 diminishes [3]. Although this fact does not deny the possibility of Li_2C_2 being formed, for example, by the reaction

$$n C (\gamma \text{- solid soln}) + m Li (L) = \frac{x}{2} Li_2C_2 (s) + (n - x) C (\gamma \text{-solid soln}) + (m - x) Li (L), \qquad (3)$$

it indicates that this will be made more difficult.

It should be noted that the results of tests carried out on carbon steels at 900 to 1000°C may be affected by the testing method used. The heating of the containers holding the samples to 900 or 1000°C and their cooling after the tests takes place over 0.5 to 1 h. Thus the carbon steels are in contact with lithium at 600 to 700°C for quite a long time, sufficient for the corrosion processes characteristic of these temperatures to take place in a surface layer up to 1 mm thick; this will alter the picture of the corrosion in the austenite phase (i.e., at the higher temperatures).

Any explanation of the corrosion mechanism in carbon steels at high temperatures, for which the carbon exists in the austenite phase, awaits further investigations.

Conclusions

1. At temperatures up to 723°C, the corrosion of carbon steels in lithium is associated with the reactive penetration of lithium into the steel as a result of interaction with cementite in accordance with Eq. (1). The lithium carbide so formed dissolves in the molten lithium surrounding the sample and the carbon content of the corrosion zone is reduced.

2. The liquid phase formed in the corrosion zone by the dissolution of Li_2C_2 in lithium tends to assist the development of the diffusion processes conveying the lithium into the steel and the removal of carbon from the steel.

3. The formation of the lithium carbide and its subsequent dissolution are accompanied by volume changes. The stresses resulting from these lead to plastic deformation of the corrosion zone.

4. The difference between the volume changes at the surface and at deeper levels of the corrosion zone leads to the development of a complex stressed state in the samples; this affects the progress of corrosion and the form of the lithium diffusion curves.

Literature Cited

1. Beskorovainyi, N. M., and Yakovlev, E. I. In collection: Metallurgy and Metallography of Pure Metals, No. 2, Moscow, Atomizdat, 1960, p. 189.
2. Beskorovainyi, N. M., et al. In collection: Metallurgy and Metallography of Pure Metals, No. 3, Moscow, Gosatomizdat, 1961, p. 233.
3. Beskorovainyi, N. M., et al. This volume, p. 121.
4. Beskorovainyi, N. M., et al. In collection: Metallurgy and Metallography of Pure Metals, No. 4, Moscow, Gosatomizdat, 1963, p. 130.
5. Grishin, V. K., et al. Properties of Lithium, Moscow, Metallurgizdat, 1963.

CORROSION OF STAINLESS CHROMIUM—NICKEL STEEL IN MOLTEN LITHIUM

N. M. Beskorovainyi, V. K. Ivanov , and V. V. Petrashko

The possibilities of using austenitic chromium-nickel steels as construction materials for use in contact with molten lithium are very limited, since these steels have an unsatisfactory corrosion resistance [1]. It is therefore of particular importance to examine the reasons for the poor corrosion resistance of chromium-nickel steels in lithium.

In this paper we shall present results of tests carried out in 1Kh18N9T steel in lithium contained in vessels made of the same type of steel or of Armco iron. The radioactive isotope C^{14} was introduced into the lithium in such quantity as to leave its total carbon content relatively unaltered; in some experiments lithium with a considerable amount of added carbon was used (Table 1).

The corrosion tests were made under static conditions at 700°C for 10 and 200 h. Then the samples were studied by the layer-analysis method. The chromium concentration was determined by a photometric flame method [2] and the nickel concentration by spectroscopic analysis.

Figure 1 shows the chromium and nickel distribution curves in 1Kh18N9T steel after tests in lithium; these indicate a reduction in the chromium and nickel content of the surface layer. In contrast to the chromium distribution curves, the nickel distribution curves have a complex character, containing a maximum and a minimum, the positions of which vary with time and test conditions.

The change in composition was least when the sample and the reaction vessel were of the same material, since the lithium saturated the chromium and nickel of the sample and vessel simultaneously (Table 2).

More considerable changes in the 1Kh18N9T steel occurred after testing in reaction vessels made of Armco iron (forms II and III). Chromium and nickel form solid solutions with iron; hence in the course of the tests chromium and nickel atoms dissolved in the lithium may pass over to the iron, and their concentration in the lithium will fall. This leads to an acceleration of the dissolution of chromium and nickel from the surface layer of the samples (see Fig. 1), which is also indicated by the weight increment of the samples: positive in tests of type I and negative in II and III (Fig. 2). Analogous results were obtained for tests on chromium stainless steels [3].

TABLE 1. Conditions for Testing 1Kh18N9T Steel in Lithium

Form of experiment	Material of reaction vessels	Added to lithium
I	1Kh18N9T	Radioactive C^{14}
II	Armco iron	Radioactive C^{14}
III	Armco iron	Radioactive C^{14} and 0.1% C (graphite)

TABLE 2. Variation in Chromium and Nickel Content at the Surface of 1Kh18N9T Steel

Form of experiment	Cr content, wt.%		Ni content, wt.%	
	after 10 h	after 200 h	after 10 h	after 200 h
I	0.9	0.9	0.8	0.7
II	0.85	0.5	0.7	0.55
III	0.8	0.5	0.6	0.5

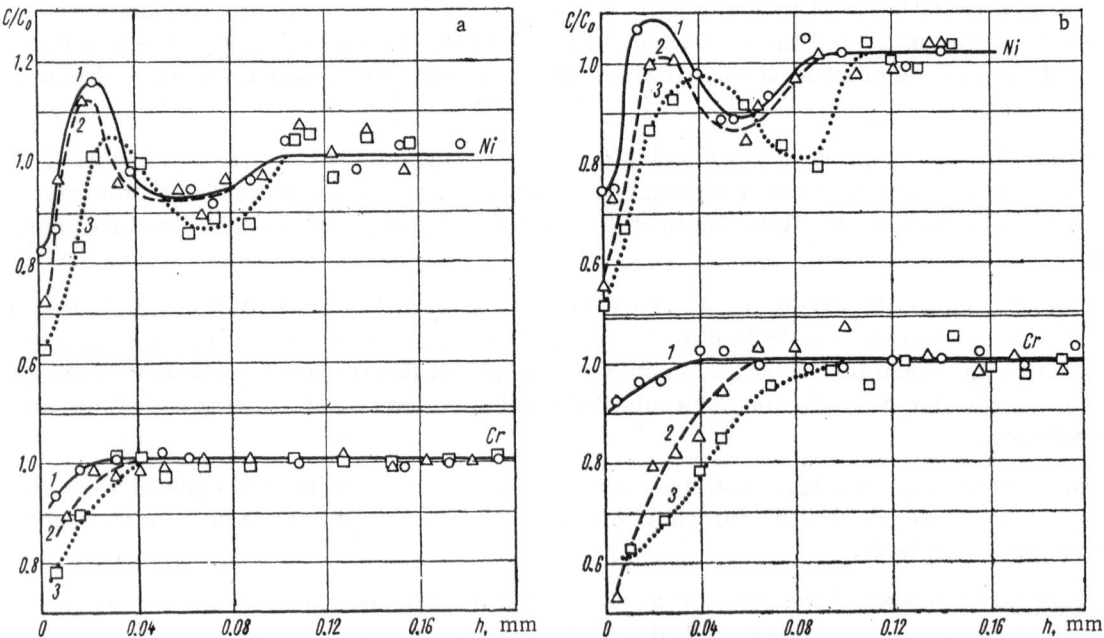

Fig. 1. Distribution of chromium and nickel in 1Kh18N9T steel after tests in lithium at 700°C for 10 h (a) and 200 h (b), using the first (1), second (2), and third (3) forms of test, respectively.

Thus test conditions II and III are more rigorous than condition I for the same type of corrosion process in 1Kh18N9T steel; because of the transfer of mass from the sample to the reaction vessel, the tests to some extent model practical conditions corresponding to the use of construction materials in molten-metal heat carriers, in which mass transfer from the hot

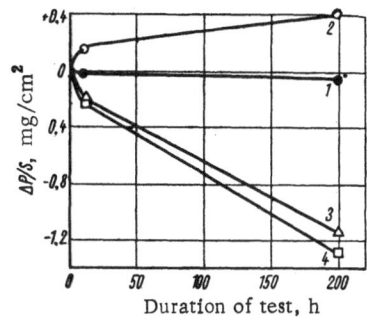

Fig. 2. Variation in the weight of 1Kh18N9T steel after tests at 700°C in argon (1) and in lithium, using the first (2), second (3), and third (4) forms of test, respectively.

zone to the cold takes place under the influence of the temperature gradient.

The results of the tests indicate the carbon has a considerable effect on the reduction in chromium and nickel content at the surface of 1Kh18N9T steel. For a higher quantity of carbon in the lithium, the corrosion processes take place more intensively.

For condition II, the amount of carbon in the lithium was approximately 0.1%, since carbon entered the lithium from the iron reaction vessels. For condition III the lithium contained around 0.2% carbon, 0.1% carbon being added in the form of graphite.

The use of radioactive C^{14} makes it possible to study the behavior of carbon during the corrosion of chromium-nickel steel in lithium. After a 10-h test under condition I (samples and vessel both made of 1Kh18N9T steel), the surface layer had a high radioactivity (Fig. 3a and Table 3) and a simultaneous considerable rise in microhardness (Fig. 4a), together with a minimal change in the composition of the surface layer with respect to chromium and nickel. This means that, for a very slight fall in the chromium content, the carbon from the lithium passes to the surface of the sample, since chromium has a higher chemical affinity for carbon than does lithium, and the microhardness of this layer probably rises as a result of carbide formation (Fig. 5).

On increasing the duration of the tests to 200 h, both the radioactivity and microhardness of the surface fall considerably (Fig. 3b and Table 3, Fig. 4b), which (in analogy with chromium steels) indicates both a certain fall in chromium content and a reduction in the amount of carbide phase on the surface [3]. Simultaneously there is a redistribution of carbon. The maximum radioactivity still remains at the surface, but there then follows a zone in which no radioactive carbon is to be found. This zone lies at the same depth as the maximum on the nickel distribution curve. At a depth of some 0.06 mm there is again a maximum on the carbon distribution curve (Fig. 3b), and this corresponds to the maximum on the curve of microhardness (Fig. 4b). At this depth the chromium concentration is approximately equal to the original, while the nickel distribution curve shows a minimum.

On contaminating the lithium with carbon (conditions II and III), the reduction in the chromium and nickel content in the surface layer of 1Kh18N9T steel takes place more intensively, but the form of the chromium and nickel distribution curves (see Fig. 1) remains the same. The carbon distribution, however, changes a great deal. On the surface, the carbon falls to very low concentrations (see Table 3). In the zone in which the nickel distribution curve has a maximum, the carbon vanishes after 10- or 200-h tests. In the zone of minimum nickel, however, the carbon concentration rises as the proportion of carbon in the lithium increases, especially after 200-h tests. The redistribution of the carbon is accompanied by a change in microhardness. At the surface, the microhardness experiences a sharp drop, while in the zone of maximum carbon content it rises.

These data indicate that, as in the tests on chromium steels, the carbon moves to deeper levels as the chromium concentration on the surface diminishes. This is due to the fact that chromium has a greater chemical affinity for carbon than have nickel, iron, and lithium.

The change in the composition of austenitic chromium-nickel steel resulting from the surface processes taking place on contact with lithium is accompanied by a change of structure.

TABLE 3. Radioactivity of the Surface of 1Kh18N9T
Steel for Various Forms of Tests, Referred
to the Radioactivity of the Sample Surface Tested
for 10 h under Condition I

Length of test, h	Experimental condition		
	I	II	III
10	1.0	0.08	0.05
200	0.16	0.04	0.02

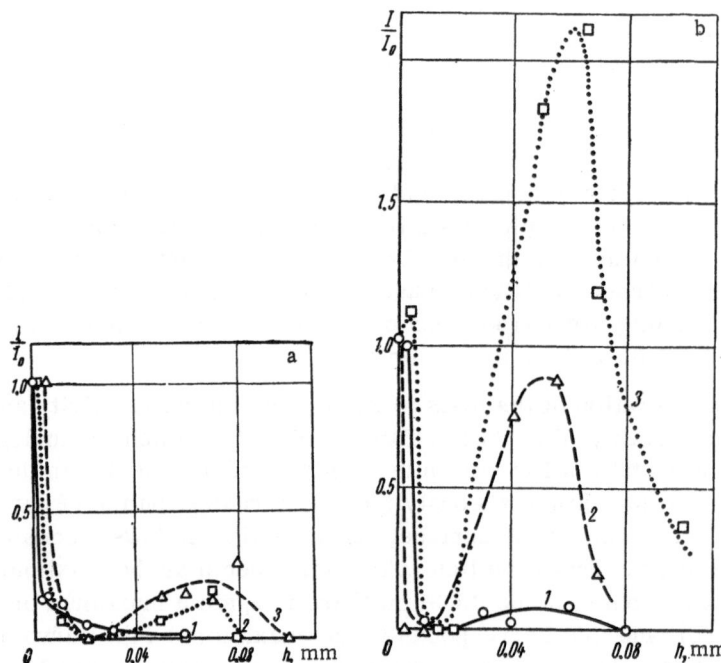

Fig. 3. Distribution of radioactive carbon in 1Kh18N9T
steel after tests in lithium containing isotope C^{14} at 700°C
for 10 h (a) and 200 h (b) by the first (1), second (2), and
third (3) test conditions.

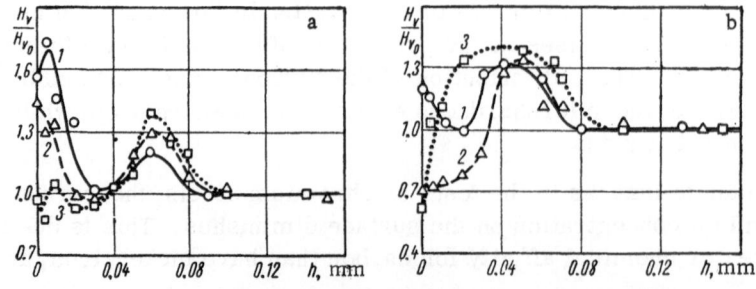

Fig. 4. Distribution of microhardness in 1Kh18N9T steel
after tests in lithium at 700°C for 10 h (a) and 200 h (b)
by the first (1), second (2), and third (3) test conditions.

TABLE 4. Change in the Mechanical Properties of 1Kh18N9T
Steel after Tests in Lithium

Test conditions	σ_B, kg/mm^2		δ, %	
	10 h	200 h	10 h	200 h
In argon atmosphere	67.4	67.8	46.2	48.2
Condition I	67.1	66.6	46.0	47.1
Condition II	69.8	69.4	44.8	26.4
Condition III	69.7	67.6	43.3	24.4

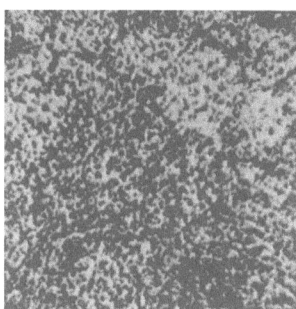

Fig. 5. Surface of 1Kh18N9T steel after tests in lithium at 700°C for 10 h, using the first test condition (× 1200).

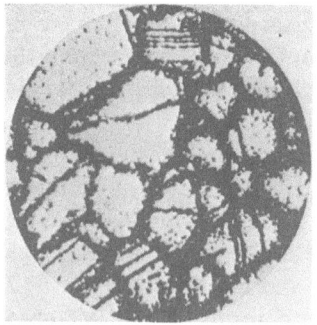

Fig. 6. Microstructure of 1Kh18N9T steel after tests in lithium at 700°C for 200 h, using the second test condition, at a depth of 0.06 mm. Etchant: 5% solution of HNO$_3$ in alcohol (× 1200).

The layers adjoining the surface contain no carbides and have an increased etchability (Fig. 6). Whereas the structure of austenitic steel is usually revealed by etching in 50% HCl + 50% H$_2$O$_2$, after tests in lithium the structure of the subsurface layers is immediately etched in a 5% solution of HNO$_3$ in alcohol, i.e., in this zone the austenitic chromium-nickel steel loses its acid-resistant properties. In the layers with increased etchability one can see pores, principally situated along the grain boundaries.

As the composition and structure of the surface layers change (tests II and III), the ductility of the 1Kh18N9T steel becomes very much lower (Table 4).

The foregoing data show that the most important process inducing corrosion damage in chromium-nickel steels on contact with lithium is the fall in the chromium and nickel content of the surface layers, accompanied by a change of structure and the development of microporosity.

On testing 1Kh18N9T steel in vessels of the same material (condition I), these processes are only weakly developed. Hence the composition and structure of the surface layer change little and the strength and ductility of the steel therefore remain similar to those prevailing after tests in argon at the same temperature.

After testing 1Kh18N9T steel in Armco-iron vessels, with the introduction of additional carbon into the lithium (conditions II and III), the surface layer loses a great deal of its chromium and nickel, and, depending on the extent to which this has taken place, the structure also changes and the ductility is lost.

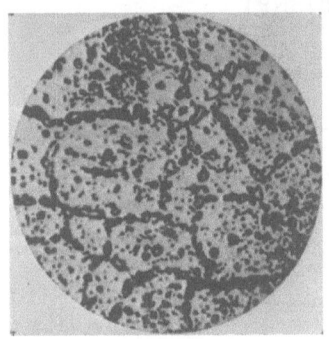

Fig. 7. Microstructure of
1Kh18N9T steel after
holding in argon for 200 h
at 700°C. Section not
etched (× 1200).

Some reduction in ductility after tests carried out under
conditions II and III may also be due to carbides formed when
the carbon diffuses from the lithium into the steel.

The main cause of the deterioration in the mechanical properties of chromium-nickel steel on contact with lithium, however,
is plainly the change in structure resulting from the loss of chromium and nickel in the surface layer. The carbon impurities in
the lithium only accentuate this process. This is indicated by the
following experiment.

Samples of 1Kh18N9T steel were tested in reaction vessels
made of Armco iron, previously decarburized in lithium. In this
case the fall in the chromium and nickel content took place more
intensively than under condition I, owing to the transfer of chromium and nickel atoms to the iron vessel, while at the same time
the effects of carbon on the corrosion process was eliminated
to the greatest possible extent. After a 10-h test, the 1Kh18N9T
steel had the following mechanical properties: $\sigma_B = 67.3$ kg /mm^2, $\delta = 47.7\%$, i.e., the ductility
was about the same as on testing under condition I. After a 200-h test the strength was $\sigma_B =$
68.1 kg/mm^2 and the ductility 31.8%. Thus after a 200-h test in a previously decarburized iron
vessel the ductility of 1Kh18N9T steel had an intermediate value between that corresponding
to condition I and that corresponding to condition II.

We may now try to explain the data obtained from tests made on 1Kh18N9T chromium-
nickel steel in lithium on the basis of the vacancy mechanism of diffusion in iron-base alloys
[4], leading to the development of pore formation and other processes accompanying this.

The dissolution of chromium and nickel atoms (and to some extent iron atoms) in lithium
should lead to the formation of vacancies at the surface of the 1Kh18N9T steel samples. The
vacancies formed partly pass out of the sample and partly diffuse inward. In the subsurface
layers, saturation with vacancies, to an extent over and above the equilibrium value determined by the composition of the alloy and the temperature, may take place. In the process
of pore formation, the alloy passes into a more stable equilibrium state [5].

The existence of a porous region in the sample (confirmed by the increased etchability
and the microstructure) may greatly change the distribution of the components with respect to
depth.

In the porous region, the diffusive segregation of the components is facilitated; for example, in the subsurface zone of 1Kh18N9T steel tested in lithium, there is an increased concentration of nickel, indicated by the maxima on the nickel distribution curves (Fig. 1).

Since nickel (in contrast to chromium) is a graphite-forming element, we find no carbon
in the zone rich in nickel atoms. The carbon diffuses into deeper layers, from which the
nickel has vanished (minima on the nickel distribution curves). In these layers there may be a
certain supersaturation of the steel with vacancies owing to the departure of the nickel atoms,
and this provides favorable conditions for the formation of chromium carbides. This is indicated by the maxima on the radioactive-carbon distribution curves and on the microhardness
characteristics (Fig. 4). On increasing the test to 200 h, the redistribution of the components
of the 1Kh18N9T steel is more substantial.

The intensification of the corrosion processes in the presence of carbon and lithium may
be due to the following causes. It is well known that chromium-nickel steels are inclined toward intercrystallite corrosion after heating in the range 500 to 800°C. This heating causes ex-

cess carbides to separate out of the γ solid solution and become distributed along the grain boundaries (Fig. 7). The diffusion of mobility of chromium is several orders of magnitude lower than that of carbon; hence the chromium required to form the carbides comes mainly from regions near the grain boundaries. Hence the regions near the grain boundaries become much poorer in chromium, and with intense carbide formation it is inevitable that vacancies should appear in these zones.

An analogous process takes place when chromium-nickel steel is in contact with lithium containing carbon impurity. The carbon, having a greater chemical affinity for chromium than for lithium, diffuses from the lithium into the steel, as indicated by the appearance of the radioactive C^{14} in the surface layers of the samples. When the carbon reaches the surface layers of the steel from the lithium, the carbide-formation process intensifies, and the regions of the γ solid solution impoverished with respect to chromium are saturated with vacancies (in addition to the vacancies formed by the dissolution of the components of the steel in the lithium).

Thus, when carbon is present in the lithium, the pore formation in the surface zones of chromium-nickel steel intensifies. On the one hand, this considerably facilitates the diffusion of chromium and nickel to the surface and their dissolution in the lithium; and on the other, it produces a sharper reduction in ductility. Finally, the carbon coming into the chromium-nickel steel from the lithium will affect the velocity of the corrosion process. It is further not impossible that the chromium atoms dissolved in the lithium may take part in the direct formation of carbides in the lithium itself. In this case the carbon impurities in the lithium should further rob the sample surface of chromium, with all the attendant consequences.

Conclusions

1. The corrosion of austenitic chromium-nickel steel on contact with lithium at 700°C is associated with a fall in the surface concentration of chromium and nickel; this produces porosity in the surface layer and reduces the ductility of the steel.

2. Carbon impurities in the lithium have a considerable effect on the degree of corrosion of 1Kh18N9T tested in lithium; they produce a more intensive fall in the surface concentration of chromium and nickel.

3. The testing of chromium-nickel steel in reaction vessels of a different material intensifies the processes taking place during the corrosion of steel in lithium.

Literature Cited

1. Beskorovainyi, N. M., et al. In collection: Metallurgy and Metallography of Pure Metals, No. 4, Moscow, Gosatomizdat, 1962, p. 122.
2. Beskorovainyi, N. M., et al. In collection: Metallurgy and Metallography of Pure Metals, No. 3, Moscow, Gosatomizdat, 1961, p. 233.
3. Nevzorov, B. A., et al. Study of the Corrosion Resistance of Construction Materials in Alkali Metals, Paper No. 343 (USSR) presented to the Third International Conference on the Peaceful Use of Atomic Energy, Geneva, 1964.
4. Krishtal, M. A. Diffusion Processes in Iron Alloys, Moscow, Metallurgizdat, 1963.
5. Geguzin, Ya. E. Macroscopic Defects in Metals, Moscow, Metallurgizdat, 1962.

APPARATUS FOR EXTRUDING METALS BY MEANS OF A HIGH-PRESSURE LIQUID

E. D. Martynov, B. I. Beresnev, D. K. Bulychev, A. N. Evstyukhin, P. Rodionov, and Yu. N. Ryabinin[*]

One of the most widespread methods of obtaining rod material and profiled components from ferrous and nonferrous metals (in addition to rolling) is the method of hot extrusion (Fig. 1a) in which the heated raw material is placed in a suitable container and extruded through a die by means of a piston.

A new method has recently been introduced for metal forming: the extrusion of metals by means of a high-pressure liquid [1-4].

The extrusion of metals in this way is accomplished by placing the original material in a container already furnished with a die of the required shape, after which liquid is forced into the container and the material is pressed through the die (Fig. 1b).

In this paper we shall briefly describe the main principles of apparatus which we have designed and used for studying extrusion processes. Early versions of the apparatus are omitted as well as apparatus still in the stage of development.

Description of Apparatus

Initially extrusion was carried out at hydrostatic pressures up to 12,000 kg/cm^2 with the apparatus shown in Fig. 2.

The apparatus consists of a container connected through a receiver to a hydrocompressor and a liquid-gas store (not shown in Fig. 2).

The liquid-gas store is intended for supplying the working liquid to the hydrocompressor at a pressure of the order of 100 kg/cm^2.

The receiver is a high-pressure vessel and serves to store high-pressure liquid in a volume much larger than that of the container.

---

[*] This work was carried out by members of the Institute of Geophysics, Academy of Sciences of the USSR, the Moscow Physical-Engineering Institute, and the Institute of Metal Physics, Academy of Sciences of the USSR, and submitted to the Scientific Conference of the Moscow Physical-Engineering Institute in 1963.

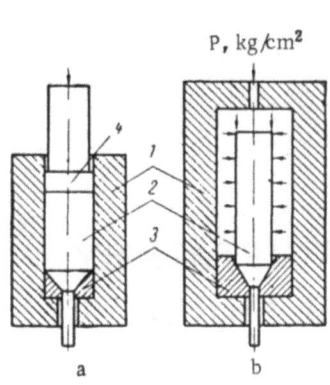

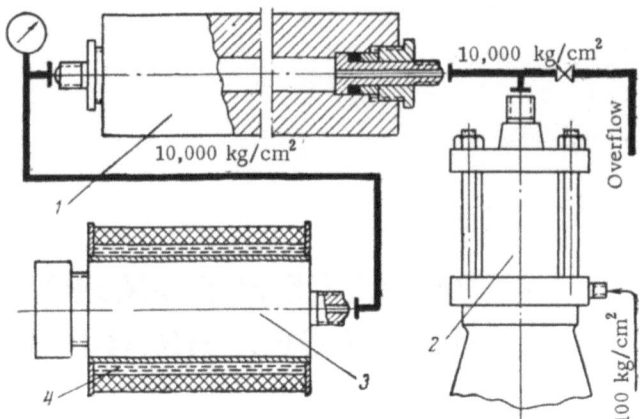

Fig. 1. Different methods of extruding metals: a) Extrusion by means of a piston; b) extrusion by means of a high-pressure liquid; 1) container; 2) original material; 3) die; 4) piston.

Fig. 2. Extrusion apparatus for pressures up to 12,000 kg/cm²: 1) Receiver; 2) hydrocompressor; 3) container; 4) electric furnace.

The pressure generator first used was a hydrocompressor of the L. F. Vereshchagin and V. E. Ivanov type [2, 5] operating up to pressures of 12,000 kg/cm². Subsequently this compressor was modernized [6], principally by changing the design of the high-pressure sealing. This modernization not only improved the reliability of the hydrocompressor but also enabled the pressure in the pressure line to be raised rather higher than 12,000 kg/cm².

Let us consider the operation of the system.

After joining the container holding the metal to the receiver, the hydrocompressor is switched on and liquid forced into the close receiver-container space. After the pressure of the liquid in this space has reached the required initial value P_i (the pressure P_i depends on the degree of deformation, the strength of the material being extruded, and the lubricating properties of the working liquid), extrusion of the material through the die commences.

The initial pressure P_i is considerably (20 to 40%) higher than the pressure P_s required for extrusion when the metal is flowing steadily. Hence, after starting the material flowing, the extrusion process continues until the pressure in the receiver-container system falls below P_s. Since the extrusion rate is very high and the pressure falls so quickly that the compressor pumps hardly any liquid through during the extrusion time, the compressor must provide a further supply of liquid at pressure P_i if the extrusion is to continue. Of course, the large volume of liquid in the receiver tends to slow the pressure drop, and if the volume of the receiver is large enough the whole of the material may be extruded without any need for supplementary recompression of liquid.

The length of material l extruded for one compression may be found approximately * from the formula

$$l = \frac{V_1 k_1 + (V_2 - V_3)\,[1 + (P_i - P_s)\,(k_2 - k_1)]\,k_2}{S\left(\dfrac{1}{P_i - P_s} - k_1\right)}\,, \tag{1}$$

*In deriving formulas (1) and (2) we ignored the elastic deformation of the receiver and container and the inertia of the extruded metal; we assumed that the liquid reaching the container from the receiver lost no temperature during extrusion.

where V_1, V_2, and V_3 are the volumes of the receiver, container, and material to be extruded, S is the cross-sectional area of the die, and k_1 and k_2 are the compressibilities of the working liquid at the temperatures of the receiver and container, respectively.*

If we wish to extrude the whole of the material in one move without extra compressions of the liquid, the receiver volume required may be found from the condition

$$V_1 \geqslant \frac{1}{k_1} \left\{ V_3 \left(\frac{1}{P_i - P_s} - k_1 \right) - (V_2 - V_3)[1 + (P_i - P_s)(k_2 - k_1)]k_2 \right\}. \tag{2}$$

If, however, the container is filled with a special working mixture differing from the liquid in the receiver, formulas (1) and (2) may still be used for practical calculations, but in this case the coefficient k_2 relates to the mixture and not the original working liquid.

The rate of flow of the metal on extrusion with this apparatus (pressure, temperature, etc. being equal), is the greater, the larger the receiver volume and the greater the compressibility of the liquid.

On increasing the degree of deformation, the velocity also rises on account of the increased pressure. In our experiments the flow velocity was of the order of 100 m/sec or higher. For receiving the extruded materials at such velocities we used felt, clay, and liquid collectors.

Extremely important for extrusion is the choice of working liquid. In the apparatus here described we used a mixture of machine oil and kerosene (70 : 30); before extrusion the container was filled with a special working medium which greatly reduced the extrusion force.

Initially extrusion was carried out at room temperature. On raising the pressure, however, the liquid and the working medium thickened and their lubricating properties worsened, which led to an increase in the extruding force and lowered the surface quality of the extruded material. In order to eliminate this defect we supplied external heating to the container by means of an electric furnace (Fig. 2). The temperature never exceeded 100 to 250°C, i.e., it was considerably below the recrystallization temperature of the majority of the metals used.

The apparatus was used for studying the principal laws governing the extrusion of metals with low yield points. On passing to stronger materials the pressure of 12,000 kg/cm² was insufficient to secure high degrees of deformation.

A considerable disadvantage of the apparatus described is the irregularity of the extrusion velocities and their inconveniently large absolute value, which often damages the extruded components. For this reason, and also to improve the reliability in operation, we set up a new system with a pressure of up to 20,000 kg/cm² in the working liquid (Fig. 3). The essential difference in this system is the use of boosters for creating the high pressure and the existence of two isolated hydraulic systems.

The low-pressure system, operating at pressures up to 1200 kg/cm², is intended for supplying the boosters and auxiliary equipment (hydraulic valve, hydraulic clamping system for the die, and so on). In this system we used standard NZhR pumps, and for smooth regulation of the pressure hand pumps of special construction giving pressures up to 2000 kg/cm² (not shown in Fig. 3). The low-pressure hydraulic system operates with machine or transformer oil.

---

*The compressibilities depend not only on temperature but also on pressure; hence we should take an average value of $P = (P_i + P_s)/2$ for the pressure.

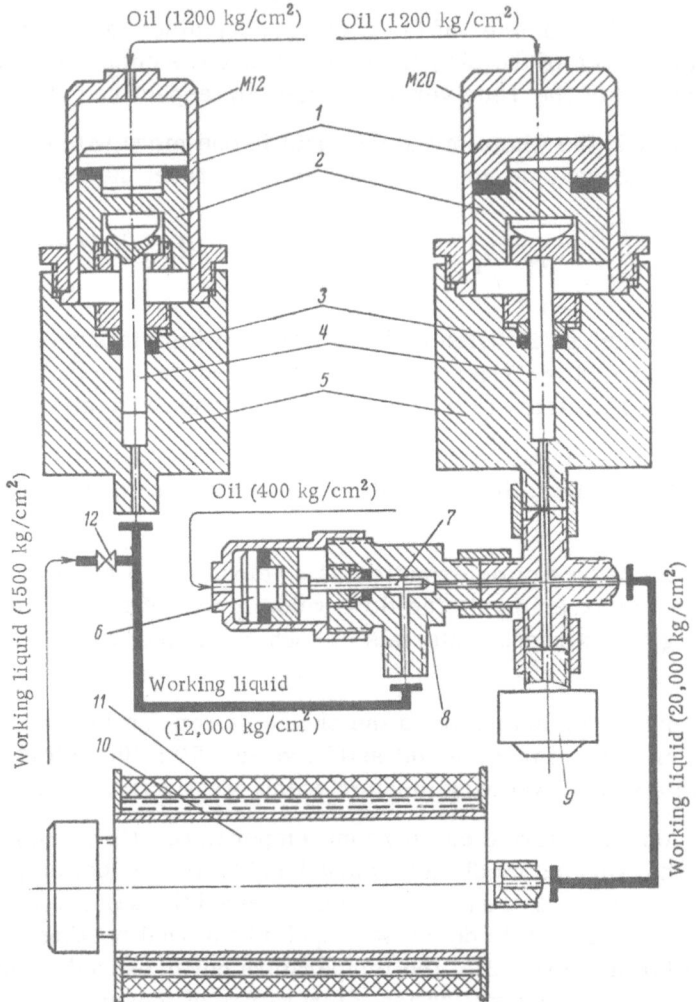

Fig. 3. Arrangement of extrusion apparatus for pressures up to 20,000 kg/cm²: 1) Low-pressure cylinders; 2, 6) pistons; 3) gaskets; 4) plungers; 5) high-pressure cylinders; 7) cut-off needle; 8) hydraulic valve; 9) manometer; 10) container; 11) electric furnace; 12) valve.

The high-pressure system operates directly on the extrusion process. High-pressure generators include M12 boosters (12,000 kg/cm²) and M20 boosters (20,000 kg/cm²), as described in [7].

When oil is fed into the low-pressure cylinder, there is a displacement of the piston, which thereupon thrusts the sealed plunger.* As a result of the motion of the plunger, the pressure of the liquid in the high-pressure cylinder and the closed system connected therewith rises. Owing to the difference in the diameters of the low-pressure piston and the plunger, the pressure in the high-pressure cylinder will be greater than that in the low-pressure cylinder. These pressures are related in the following way:

$$P_1 = P_2 k_{\text{B}}, \tag{3}$$

_____
* The sealing of the plunger operates on the principle of the uncompensated area and has a 25% degree of noncompensation.

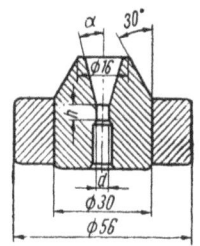

Fig. 4. Arrange-
ment of die.

where P_1 and P_2 are the pressures in the high and low-pressure cylinders, respectively, and k_B is the boosting factor.

The factor k_B is given by the formula

$$k_B = \frac{d_2^2}{d_1^2} - \Delta f, \tag{4}$$

where d_1 and d_2 are the diameters of the plunger and low-pressure piston, respectively, and Δf is a correction for friction in the gaskets (determined experimentally). Usually the value of Δf varies over the range $\Delta f = (0.03-0.1)d_2^2/d_1^2$, depending on the construction of the gaskets. In the new system the boosting factors were $k_B = 10$ for the M12 booster and 17 for the M20.

Let us consider the operating principles of the equipment.

After placing the material to be extruded in the container, the working liquid is forced through a cut-off valve into the high-pressure system. The high-pressure cylinders and container are thereupon filled with liquid, and the plungers of the boosters are raised to the upper working position.

On reaching a pressure of 1500 to 2000 kg/cm² the valve shuts off and the forcing of oil into the low-pressure cylinder of the M12 booster begins. As a result of the oil pressure the piston thrusts the plunger, which produces a corresponding rise in pressure over the whole high-pressure system. After the pressure in the high-pressure system reaches 12,000 kg/cm², the operation of the M12 stops, and the oil is fed to the cylinder of the hydraulic valve. Under the pressure of the oil, the piston of the hydraulic valve moves the cut-off needle, so that the M12 is cut off from the rest of the high-pressure system.

Then oil is forced into the low-pressure cylinder of the M20 booster, as a result of which the pressure in the high-pressure system rises to 20,000 kg/cm². After the pressure has reached the required P_i, extrusion of the material begins.

If, as a result of the large volume of the container, one cycle of the M12 booster is insufficient to produce 12,000 kg/cm² over the whole high-pressure system, then after closing the hydraulic valve the oil must be driven out of the M12 cylinder, and, after opening the valve, the working liquid must again be pumped into the M12. Immediately afterwards the valve is closed, the hydraulic valve opened, and oil is again forced into the cylinder of the M12 booster.

This cycle is repeated until the whole system reaches the pressure of 12,000 kg/cm²; after this the operation of the M20 booster begins.

If one cycle of the M20 is insufficient to extrude the whole amount of material in the container, the entire cycle must be repeated for both boosters.

When working with a container of large volume, a second hydraulic valve is set between the M20 and the container. In this case, when repeating the cycles and pumping fresh quantities of working liquid into the high-pressure system, the container can be shut off and retain its high pressure; this considerably increases the efficiency of the apparatus. Naturally, if the extrusion of all the material requires a pressure below 12,000 kg/cm², we may use simply the M12 booster with an open hydraulic valve.

Since this apparatus does not contain a receiver, and the volume of the working liquid in the system is therefore fairly small, the rate of metal outflow can be controlled very smoothly by varying the supply of oil to the low-pressure cylinders of the boosters.

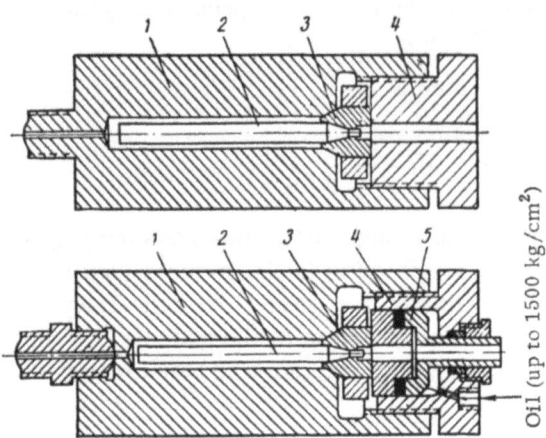

Oil (up to 1500 kg/cm²)

Fig. 5. Arrangement of containers: Top) Container for pressures up to 12,000 kg/cm²; below) container with hydraulic die for pressures up to 20,000 kg/cm². 1) Body; 2) material ready for extrusion; 3) die; 4) clamping screw; 5) piston.

In our experiments we were able to regulate the outflow velocity (in the steady condition) between a few millimeters and several meters per second. In this way we can stop the extrusion process at any stage and thus avoid the "shooting" of the extruded material at the end of the flow period.

It was not possible to use the working liquid mentioned earlier (machine oil and kerosene) in the high-pressure system, since this thickens at pressures of 15,000 kg/cm² and over. The working liquid used was therefore a mixture of glycerin and ethyleneglycol (60:40), which changes its viscosity very little up to pressures of 30,000 kg/cm².

In order to reduce friction and the extruding force, special working media were used in the new system (as in the earlier 12,000 kg/cm² system); these media, insoluble in the working liquid, were applied to the material to be extruded as a thin layer of paste. The lubricating properties of such pastes, however, worsen at high pressures. For this reason, and also on principles associated with the plastic flow of various metals, the container was heated by means of a tubular electric furnace (Fig. 3). The temperature was kept between 100 and 350°C, or appreciably higher in certain cases.

It should be noted that the mixture of ethyleneglycol and glycerin used as working liquid at pressures around 2000 kg/cm² was neither decomposed nor oxidized at 350 to 400°C. Separate experiments suggested that at pressures above 10,000 kg/cm² the temperature of this liquid could rise above 500 or 600°C.

The high pressure in the working liquid is measured by means of an electrical-resistance manometer (Fig. 3). It proved convenient to plot calibration curves for each of the boosters in $P_1 - P_2$ coordinates from the readings of this manometer, and by using these graphs to determine the pressure in the working liquid from the readings of manometers in the low-pressure system.

Dies and Containers

The most frequently-used die is illustrated in Fig. 4.*

The die consists of a body pressed into a tire ring. The outer jaw of the die serves to clamp the die into the container. This is effected by tightening the jaw of the die into a corresponding conical expansion of the container channel (Fig. 5). A contact stress σ_c develops at the junction between the conical surfaces. If this stress is greater than the maximum pressure of the working liquid inside the container, no leakage of liquid takes place.

In practice a ratio of $\sigma_c \approx 1.2 P_i$ is required. In this case the necessary tightening stress is given by the formula

$$F \approx P_i \cdot \frac{\pi}{4} \left(1.2 D^2 - 0.2 d_1^2 \right), \tag{5}$$

* The dimensions in Fig. 4 correspond to a die for a container of internal diameter $d_1 = 16$ to 18 mm.

where F is the stress, D the maximum diameter of the conical expansion in the container channel, and d_1 the internal diameter of the container.

The length h of the gaging part of the die is made within limits of h = (0.5–1.0)d, where d is the diameter of the gaging part. For h = 0.5d the extrusion force is rather smaller (by 5 to 10%) but the material extruded may be bent (undulating extrusion).

The conical angle of the going-down part of the die greatly affects the extrusion force and the quality of the resultant material. We used dies with various angles of entrance cone (between α = 2 and 45°).

The tire ring serves to increase the strength of the body of the die.

The clearance for pressing the body of the die into the tire ring is 0.5 to 0.8% of the inner diameter of the ring, so that the ring automatically clamps itself in the course of press fitting. We may note that for pressures up to 10,000 kg/cm² dies with central openings of simple shape (no stress concentrators) need no tire rings.

The body of the die is made of ShKh15 steel (R_C = 57 to 60) or ÉI643 (R_C = 53 to 55), and for extrusion temperatures above 200°C of 3Kh2V8 steel (R_C = 53 to 55).

The material for the tire ring is 45KhNMFA steel (R_C = 40 to 44).

The die construction described proved entirely satisfactory up to a pressure of 12,000 kg/cm² inside the container at the preliminary tightening stress, and up to 20,000 kg/cm² with a variable tightening stress, rising in proportion to the pressure in the container.

The container itself is one of the most important components of the extrusion system. Typical constructions of our containers are indicated in Fig. 5.

The container shown in Fig. 5a was used for pressures up to 12,000 kg/cm². It consisted of a body and an adjusting screw. The adjusting screw serves to clamp the die with the extrusion material in it to the conical expansion of the container channel. The screw is tightened by hand, using a long wrench, which is rather troublesome in operation, especially at high extrusion temperatures.

For working-liquid pressures above 12,000 kg/cm² in the container and also for large diameters of container channel, hand tightening of the screw is insufficient to fix the die adequately. Both for this reason and for convenience in operation, a revised form with a hydraulically-tightened die (Fig. 5b) was constructed. The new form had the distinguishing feature that a special hydraulic system was installed in place of the screw; the hydraulic system consisted of a cylinder and tightly fitting piston.

After placing the die and the extrusion material in the container channel, the clamping system is screwed into the container. The force required to clamp the die is created by forcing oil into the cylinder of the hydraulic clamping system. In order to fix the die for a pressure of 20,000 kg/cm² in the container channel, the oil in the cylinder of the clamping system must have a pressure of 1400 to 1500 kg/cm². The piston of the clamping system has an opening for letting out the extruded material.

An important aspect of the construction described is the ability to vary the die–clamping force as the pressure of the working liquid in the container varies, which is especially important for increasing the strength of the die.

The body of the container is made as a self-reinforced single layer. The materials for this include 40Kh, 45KhNMFA, and ÉI643 steels, and for extrusion temperatures above 250°C 3Kh2V8.

Design of High-Pressure Vessels

Such important components of extrusion equipment as the containers, booster cylinders, and receivers constitute high-pressure vessels. Let us consider the design and construction of these in more detail.

The high-pressure vessels may be single- or multilayer and operate under elastic, elastic-plastic, or entirely plastic conditions.

Calculations for the strength of thick-walled vessels have been developed in full detail in a number of papers [8-10].

Let us call r_1 and r_2 the inner and outer radii of the vessel, respectively, and denote their ratio by $n_2 = r_2/r_1$.

According to [8], the maximum internal pressure for a single-layer closed cylindrical vessel operating in the elastic condition is given by the formula*

$$P = \frac{\sigma_s \left(n_2^2 - 1\right)}{n_2^2 \sqrt{3}}, \tag{6}$$

where σ_S is the yield stress of the vessel material.

We see from expression (6) that with increasing wall thickness of a vessel operating in the elastic condition, the maximum possible internal pressure increases more slowly, and even for a vessel with an infinitely thick wall the value of this pressure never exceeds $\sigma_S/\sqrt{3}$. This is because the stresses in the wall of the vessel are distributed nonuniformly and, after reaching a maximum at the inner surface of the vessel, rapidly fall off over the wall thickness. A considerable part of the vessel wall is thus unstressed. Thus, if the vessel is used in the elastic condition, merely increasing the wall thickness will not allow the maximum pressure to be raised to 10,000 or 12,000 kg/cm^2, even if the strongest materials are used.

In order to raise the permissible internal pressure, we may use multilayer reinforced vessels with prestressed walls. The reinforced vessels consist of two or more cylinders pressed tightly one against the other. This produces prestressing in the walls, and gives a more uniform distribution of stresses over the wall thickness when the vessel is subjected to internal pressure; the maximum internal pressure permissible is thus increased. Reinforced vessels are usually two-layered.

Let us call the radius of the layer interface in the two-layer vessel r_c and the inner and outer radii r_1 and r_2, respectively. If the fitting of the outer layer is carried out correctly,† the two layers should be equally strong, i.e., on applying an internal pressure to the vessel the two layers should exhaust their reserve of elastic resistance simultaneously.

The maximum internal pressure for a two-layer vessel reinforced optimally may be determined from the formula

$$P = \frac{\sigma_s}{\sqrt{3}} \left[2 - \left(\frac{r_1}{r_c}\right)^2 - \left(\frac{r_c}{r_2}\right)^2 \right]. \tag{7}$$

Differentiating expression (7) with respect to r_c, we obtain the conditions for maximum P in the form

$$r_c = \sqrt{r_1 \cdot r_2}. \tag{8}$$

---

* The vessel-design formulas are based on the Mises-Hencky plasticity conditions.
† Directions for finding the optimum fit are given in more detail in [8].

TABLE 1. Calculated Internal Pressure for Single-Layer
Vessels

$n_2 = \dfrac{r_2}{r_1}$	P_1, kg/cm²	P_2, kg/cm²	P_s, kg/cm²	P_h, kg/cm²
2	3900	5200	7200	9500
4	4870	7800	14400	18100
6	5050	8660	18600	24300
8	5120	9100	21600	28800
10	5150	9360	24000	31400
∞	5200	—	—	—

Condition (7), called Gadolin's relation, enables us to find the most advantageous radius r_c for given internal and external radii of the vessel.

A considerable rise in the maximum internal pressure may be achieved by self-reinforcing (self-loading) of single-layer vessels. Self-reinforcing means plastic deformation of the vessel under the influence of internal pressure, as a result of which there is a redistribution of stresses over the wall thickness, and the wall becomes more evenly stressed than under the elastic loading of single- or multilayer vessels. Self-reinforcement may be carried out either preliminarily, with subsequent machining of the vessel channel, or directly in the course of actual operation, if there is no need for exactitude in the channel dimensions.

For a practical determination of the maximum internal pressure in a self-reinforced vessel, we may use the following formula [8]:

$$P = \frac{\sigma_s}{n_2^2 \sqrt{3}} \left(2 \ln n_0 + n_2^2 - n_0^2\right),\tag{9}$$

where $n_2 = r_2/r_1$ and $n_0 = r_0/r_1$, and r_0 is the radius of the plastic zone.

For complete self-reinforcement, when the plastic zone extends over the whole wall thickness ($r_0 = r_2$), expression (9) takes the form

$$P = \frac{2\sigma_s}{\sqrt{3}} \ln n_0.\tag{10}$$

An important fact is that the material of the vessel walls is hardened by the plastic deformation, so that formulas (9) and (10) give rather too low values of the maximum pressure.

A method of calculating self-reinforced vessels, allowing for the hardening of the material, is set out in [8-10].

In order to illustrate the above discussion, we calculated the maximum internal pressure for various relative wall thicknesses in single- and double-layered vessels operating under elastic conditions, and also for self-reinforced vessels. The vessel material was 33KhN3MA steel with a yield point of $\sigma_s = 90$ kg/mm². The results of the calculations appear in Table 1, where P_1 and P_2 are the internal pressures for single-layer vessels operating in the elastic conditions and double-layer vessels with optimum reinforcement as given by the Gadolin equation; P_s denotes the pressure for completely self-reinforced vessels, calculated without allowing for hardening, and P_h is the pressure for a completely self-reinforced vessel after making due allowance for hardening.

We see from Table 1 that self-reinforcement is a more effective means of increasing the maximum internal pressure than the use of reinforced vessels. In addition to this, self-reinforced vessels are simpler and cheaper to make than multilayer types.

TABLE 2. Calculated Internal Pressure
for Two-Layered Vessels

P_S, kg/cm^2	$n_0 = \dfrac{r_0}{r_1}$	$\dfrac{\Delta d_1}{d_1}$, %
25000	5.0	11.8
24600	4.29	9.25
24000	3.93	7.35
23200	3.4	5.55
22000	3.0	4.3
21300	2.78	3.7
19800	2.45	3.0
18900	2.29	2.6

It should be noted that, in order to effect complete self-reinforcement, considerable plastic deformation is required in the walls of the vessel. Hence in vessels with large relative wall thickness ($n_2 \geq 5$) it may happen that the reserve of plasticity in the material will not cope with complete self-reinforcement and the localization of deformation, so that the vessel will suffer rupture before the whole wall passes into the plastic state. Usually, however, there is no need for full self-reinforcement.

Let us consider as an example a vessel with an outer-to-inner radius ratio of $n_2 = 5$, made of 45KhNMFA steel with a yield point of $\sigma_S = 135$ kg/mm^2. The maximum internal pressure for the elastic operating condition of this vessel is 7500 kg/cm^2, or if the vessel is made two-layered (satisfying the Gadolin criterion), 12,500 kg/cm^2.

Table 2 gives the internal pressure for a vessel of this kind after self-reinforcement, for various degrees of plastic deformation. The first column of Table 2 gives the pressure inside the self-reinforced vessel, calculated without allowing for hardening, the second column gives the corresponding values of $n_0 = r_0/r_1$ giving the extent of the plastic zone in the walls of the vessel. The third column indicates the corresponding increments to the internal diameter of the vessel resulting from the plastic deformation.

We see from the data of Table 2 that even incomplete self-reinforcement gives quite a good effect.

In practice, for vessels with $n_2 = 5$ to 10, we usually limited the degree of self-reinforcement to that giving a 4 to 7% increment in internal diameter.

It was shown in [8] that in self-reinforced vessels with $n_2 > 2.22$, secondary plastic deformations near the inner surface may occur on unloading. From this we might well suppose that with a fairly large number of loadings and unloadings the wall material would be softened as a result of the Bauschinger effect.

In this case, owing to the softening of the vessel walls, the plastic deformations would grow with each new loading, and after a certain number of cycles of loading and unloading the vessel would break up. It was shown experimentally in [11], however, that, if a self-reinforced vessel is repeatedly loaded with dynamic pressures not exceeding a certain critical value, then the plastic deformations die out asymptotically with increasing number of loadings.

Our own measurements of the internal channels of vessels on loading with static pressure showed an analogous damping of the plastic deformations after 8 to 12 loading-and-unloading cycles, after which the vessel withstood an unlimited number of loadings without serious deformation. There was no case of failure in vessels which had been subjected to correct heat treatment, even after many hundreds of loadings.

The nature of these phenomena has not yet been studied very greatly, but we may suppose from Bridgman's data [1] that the high pressure inside the vessel to some extent weakens the Bauschinger effect.

It should be noted that what has been said does not apply to vessels with stress concentrators in the form of sharp edges or long grooves. Such vessels may sometimes fail after a certain number of loadings in a manner resembling fatigue failure.

One of the main factors determining the strength of a self-reinforced vessel is the correct choice of material and heat treatment. Thus, for example, the complete self-reinforcement of vessels (especially for $n_2 > 4$) is impossible for many high-strength steels because of their inadequate ductility.

Steels used for making self-reinforced vessels for pressures above 10,000 kg/cm^2 should have the following properties: yield stress $\sigma_s \geq 90$ kg/mm^2 (preferably above 130 kg/mm^2), relative elongation at rupture $\delta \geq 8\%$ (preferably above 10%), relative contraction $\psi \geq 40\%$, impact strength $a_H \geq 4$ kg · m/cm^2.

In order to obtain the required properties, very strict conditions must be imposed on the heat treatment of high-pressure vessels. We ourselves had a case in which, out of a batch of 30 vessels with a nominal hardness of $R_C = 42$ to 44 (45KhNMFA steel), five had a hardness of $R_C = 46$ to 47 after heat treatment. Under tests all five failed at a pressure of 22,000 kg/cm^2, whereas not one of those 25 subjected to correct heat treatment failed.

It is quite impermissible to quench high-pressure vessels to the maximum values of ultimate-tensile stress for the material in question.

Let us use 40Kh steel as an example of the effect of heat treatment on the strength of the vessel. Steel of this type tempered at 150°C has the following properties: $\sigma_s = 130$ kg/mm^2, $\delta \approx 1\%$. For this tempering temperature, the maximum calculated pressure for a vessel with $n_2 = 5$ is 7200 kg/cm^2, since owing to the inadequate reserve of plasticity failure ensues almost immediately after the elastic state of the cylinder has been passed (actually such vessels ruptured at a pressure of 7000 to 9000 kg/cm^2).

If the same vessel is tempered at 470 to 500°C, the yield stress of the 40Kh steel falls to $\sigma_s = 90$ kg/mm^2, while the relative elongation rises to $\delta = 12\%$. In this case full self-reinforcement is possible, and the maximum internal pressure, according to formula (10), reaches 16,800 kg/cm^2.

Thus under correct heat treatment the strength of the vessel more than doubled (without allowing for hardening), despite the considerable fall in the yield stress of the material (in practice such vessels have been used for long periods at a pressure between 14,000 and 15,000 kg/cm^2).

A no less important factor in selecting a material for the vessel is its hardenability. Thus some steels cannot be used for high-pressure vessels in spite of having the recommended combination of strength and ductility, because of their poor hardenability.

The materials which in our view are the most promising for high-pressure vessels include ÉI643, 45KhNMFA, and 15Kh2GN2TRA steels, and for vessel-wall thicknesses above 100 to 120 mm types 33KhN3MA and 30KhGSNA. For vessels working at 300 to 500°C, 3Kh2V8, 40KhNMA, 23Kh2NVFA, etc., steels may be used.

If as a result of too great wall thicknesses the optimum strengths cannot be achieved by appropriate heat treatment, then this fact must be taken into consideration when calculating the strength of the vessel by putting the actual values of σ_s into the design formulas.

If the high-pressure vessels are used at high temperatures, the reduction in the strength of the material due to this must also be considered.

The calculation of vessels with a uniform temperature distribution may be carried out by substituting the values of σ_s corresponding to the given temperature in the design formulas. The calculation of vessels heated nonuniformly is set out in [9].

Literature Cited

1. Bridgman, P. V. Study of Large Plastic Flow and Fracture [Russian translation], Moscow, IL, 1955. [English edition: McGraw-Hill, New York, 1952.]
2. Beresnev, B. I., et al. Some Questions Relating to Large Plastic Deformations of Metals at High Temperatures, Moscow, Izd. Akad. Nauk SSSR, 1960.
3. Beresnev, B. I., et al. Fiz. Metal. i Metalloved., 11:(1961).
4. New Scientist, 18:333 (1963).
5. Vereshchagin, L. F., and Ivanov, V. E. Author's Certificate No. 661,990, January 24, 1958.
6. Beresnev, B. I., and Ivkov, V. P. Pribory i Tekhn. Éksperim., No. 5:(1961).
7. Livshits, L. D., and Martynov, E. D. Pribory i Tekhn. Éksperim., No. 3:(1963).
8. Belyaev, N. M. Theory of Elasticity and Plasticity, Moscow, Gostekhizdat, 1957.
9. Il'yushin, A. A., and Ogibalov, P. M. Elastic-Plastic Deformation of Hollow Cylinders, Moscow, Izd. MGU, 1960.
10. Lomakin, V. A. Large Deformations of a Tube and Hollow Sphere, Ind. Sb. Akad. Nauk SSSR, 1955, Vol. 21.
11. Ogibalov, P. M. Izv. Akad. Nauk SSSR, Otd. Tekhn. Nauk, No. 9:(1958).

APPARATUS FOR MEASURING HOT HARDNESS
BY STATIC METHODS

Yu. G. Godin, N. A. Evstyukhin,
A. A. Kul'bakh, and V. M. Shchavelin

Introduction

The measurement of hardness is one of the most important methods in the physicochemical analysis of metallic and nonmetallic systems. This method is frequently used for estimating the strength and ductility of metals and alloys, since the hardness is related to the tensile strength and other mechanical characteristics. In hardness tests the sample is not taken up to the rupture point and hence the state of the grain boundaries does not affect the hardness readings.

Of special significance is the measurement of hot hardness in estimating the properties of refractory and brittle materials (carbides, nitrides, borides, etc.), from which it is hard to prepare rupture-test samples owing to their brittle properties.

Presently known methods of hot hardness testing are divided into static and dynamic [1]. The most widespread are static methods of determining hardness by means of a constant indentor (sphere, cone, pyramid). The most accurate data are obtained by measuring hardness with a diamond (or sapphire) pyramid having an angle of 136° between opposite faces, since in this case the hardness depends very little on the load, and the cold-working of the material is less than with tests involving conical tips.

On measuring hardness with an indentor in the form of a steel sphere, similarity of the impressions is not preserved, so that the hardness depends greatly on the applied load. Hardness may be measured at temperatures up to 1750°C with a pyramidal sapphire indentor.

According to [2], by combining the method based on the static impression of a sapphire indentor with that of the unilateral flattening of a conical sample with a cone angle of 120°, the temperature threshold of hardness measurements may be raised to 3000°C. For measuring hardness at high temperatures, a system based on the dynamic method of the rebounding sphere has been proposed [3].

The arrangements proposed in these various sources for measuring hot hardness at high temperatures nevertheless have some disadvantages. For example, in the apparatus for measuring hot hardness by static methods proposed in [2], these include: 1) difficulty of automating the loading of the sample and conveying it to the heating zone; 2) impossibility of ensuring uniform heating of the sample because of the design of the heater; 3) complexity and laborious nature of the load-calibrating system; 4) complexity of adjusting the apparatus during measure-

YU. G. GODIN ET AL.

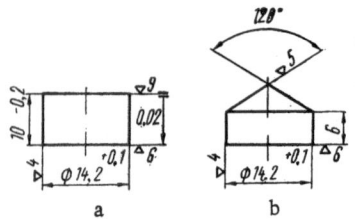

Fig. 1. Samples for studying the hardness of metals
and alloys at high temperatures: a) By the impres-
sion method; b) by unilateral
flattening.

ment, etc. These disadvantages affect the accuracy of hardness measurements at high temperatures. In this paper we
shall describe an improved system for measuring high-temperature hardness, entirely free from these failings.

Principle of the Apparatus

The apparatus is intended for measuring the hardness of
metals and alloys over a wide temperature range, from room
temperature to 3000°C. The basis of measurement is the
scheme developed by V. A. Borisenko in the Institute of Metallo-
ceramics and Special Alloys, Academy of Sciences of the Ukr.
SSR [2].

According to this scheme, the hardness of samples is
measured by two methods:

1) At temperatures up to 1500°C, by the static impression of an indentor of the standard
tetrahedral-pyramid type with an angle of 136° between opposite faces (the sample tested has
the shape and size indicated in Fig. 1a for this case);

2) at temperatures above 1500°C, by the unilateral flattening of conical samples with a
vertical angle of 120° (shape and size indicated in Fig. 1b). The flattening of the sample is effected by means of a plane piston or punch, which takes the place of the indentor. A measure
of the sample hardness is the ratio of the load to the area of the crushed vertex of the cone.

Figure 2 shows the principle of the apparatus. Samples 2 and 3 are placed in sockets of
the interchangeable stands 1, which can be attached to the upper end of the rod 12.

Rod 12 has two degrees of freedom:

1) To-and-fro motion in directions A for introducing the samples into the hot zone, subjecting to the load, and returning to the original lower position:

2) rotational motion relative to the vertical axis, so as to be able to apply the load at a
number of points on the circumferences of circles of various radius.

The to-and-fro motion of the rod is produced by an electric motor through worm coupling
15 and drive 17. Control of this motion is semi-automatic. The rod is rotated manually by a
pilot wheel mounted on the control desk. The rotation of the rod causes rotation of the cylinder
13 through the worm coupling 14; the cylinder is linked to the rod by the pivot 16.

The loading unit is completely housed in the working chamber. Either an indentor 5 or a
plunger 4 can be fixed to the load slide 9.

The sliding rod 9 has two degrees of freedom:

1) To-and-fro motion in the B directions (the working, or loading and unloading, motion);

2) to-and-fro motion in the horizontal plane in the C direction, enabling the axis of the
indentor or plunger to be moved up to 6 mm from the axis of the rod 12.

Ten 1-kg loads are lined up on the load rod 9 before the start of the experiment. Twenty-
six samples are loaded into the magazine 10, from which they can be pushed individually into
the socket of the stand 1 by moving the rod 11.

Using the method based on the impression of a sapphire indentor, 40 impressions may be
made on one sample; using the unilateral-flattening method, up to 26 measurements may be
made.

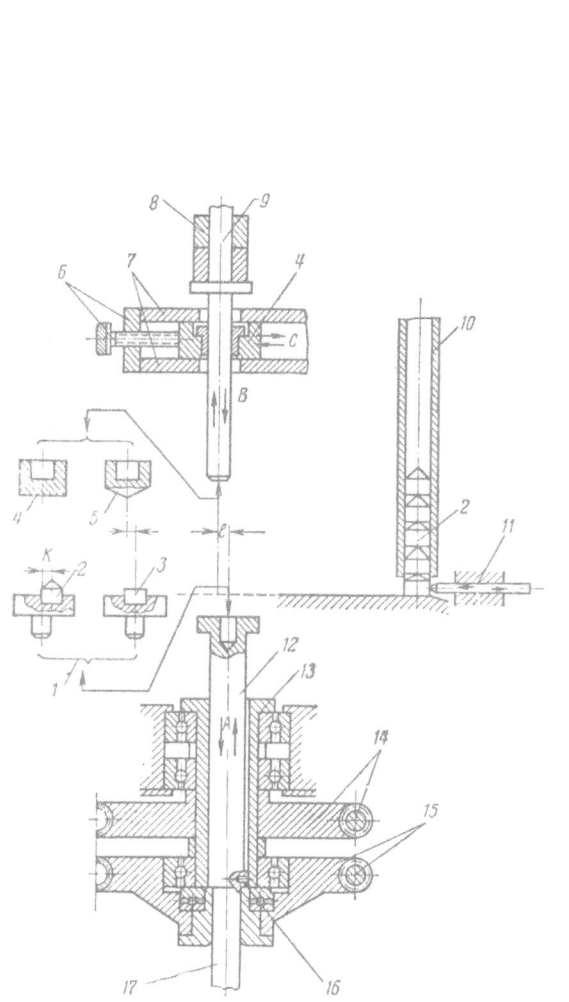

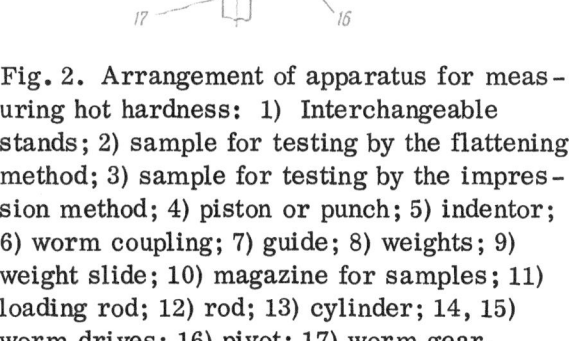

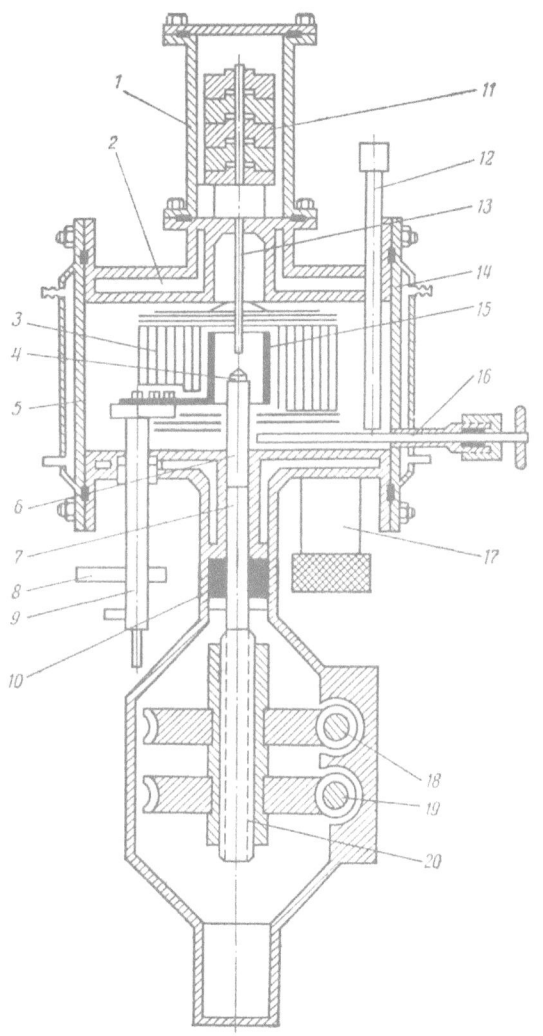

Fig. 2. Arrangement of apparatus for measuring hot hardness: 1) Interchangeable stands; 2) sample for testing by the flattening method; 3) sample for testing by the impression method; 4) piston or punch; 5) indentor; 6) worm coupling; 7) guide; 8) weights; 9) weight slide; 10) magazine for samples; 11) loading rod; 12) rod; 13) cylinder; 14, 15) worm drives; 16) pivot; 17) worm gear.

Fig. 3. Main component of the apparatus for measuring hot hardness: 1) Cylinder; 2) body of chamber; 3) screens; 4) sample; 5) side flange; 6) stand; 7) rod; 8) clips; 9) current leads; 10) vacuum seal; 11) loads; 12) sample magazine; 13) load rod; 14) side flange; 15) heater; 16) loading device; 17) sample collector; 18, 19) worm drives; 20) worm coupling.

The apparatus is designed for working either in vacuum or in a medium of purified inert gases.

Construction of the Apparatus

The main part of the apparatus (Fig. 3) consists of the following units: body of the chamber 2, a lifting and rotating stand 6, loading device 16, indentor unit 11, 13, heater 15 with leads 9, and semi-automatic electric-motor control system.

Body of the Chamber. The cylindrical body of the chamber 2 has double water-cooled walls, the water being fed in through a conduit welded into the chamber. The ends of the

Fig. 4. Heater.

chamber are closed by removal flanges 5 and 14 (vacuum-tight). Each flange has its own water jacket and is fixed to the body by means of bolts. A guide bushing for the loader rod is welded into the flange 14.

At the top, the body has a flange for attaching the indentor unit 11, 13 and the loading magazine for the samples 12. At the bottom, welded to the base, is a tube containing the mechanism for the lifting and rotating stand 6, incorporating both bearings and drive. Two electrical leads 9 pass through the base plate, and a tube 17 for collecting the samples is also welded to the base.

Two tubes are welded into the sides of the body: One of these is closed with vacuum-tight glass for pyrometric control of the temperature, and the other is connected to an oil-vapor TsVL-100 diffusion pump. The second tube is closed with a vacuum-valve and has two inlets intended, respectively, for connecting vacuum gages and filling the chamber with an inert gas. The base plate of the body has bracing clamps for mounting the vacuum chamber on the apparatus stand.

Lifting and Rotating Table. The lifting and rotating stand 6 is connected to the vertical rod 7 which derives its up-and-down motion from the worm coupling 20 and worm drive 19. The worm 19 is set in rotation by the electric motor through a friction safety clutch. The DVS-U1 electric motor (tape-recorder type, synchronous, reversing) has a power of 30 W and a velocity of 1500 rpm. Rotation of the stand relative to the vertical axis is effected by means of a pilot wheel on the control desk. The wheel imparts rotation to the stand through worm coupling 18. On the control desk is a revolution counter for the pilot wheel. Inside the stand, rod, and screw is a series of openings for the introduction of thermocouples.

Indentor Unit. The indentor unit is situated entirely within the vacuum chamber and is fixed to the upper flange. The indentor is placed in a mounting fixed to the central bushing, which in turn is able to move freely in a vertical direction. This motion effects the loading of the samples in testing. It is also possible to move the bushing with the indentor in a horizontal direction by means of a worm coupling by rotating a shaft with a graduated circle introduced through a bellows from the vacuum chamber. This motion, in combination with the rotation of the stand around its own axis, enables impressions to be made at different points of the samples (or the sample to be flattened at a different point on the surface of the plunger every time). At the top, on a pin centered with respect to the mounting, are interchangeable weights 11 (maximum number of weights: 10 at 1 kg each).

The load can be varied from 0.5 to 10 kg at intervals of 0.5 kg; it is set before the experiment and cannot be varied in the process of testing.

Heater and Supply System. The sample is heated by radiation from a tungsten heater (Fig. 4). The heater has an "omega" shape with openings for pyrometric temperature control (height 80 mm, diameter 35 mm). This heater construction enables the samples to be heated uniformly to high temperatures. The heater is fed from an OSU-40/0.5 transformer situated within the apparatus table. In order to obtain smooth temperature control, a single-phase AOSK-25/0.5 autotransformer with a nominal power of 40 KVA is employed. The current and voltage in the circuit are measured by control systems set at the back of the panel.

Electrical Circuit of the Semi-Automatic Control System. The feeding of the sample into the heating zone, the loading, unloading, and holding, and the release of the sample from the heating zone are effected semi-automatically by means of the electrical circuit shown in Fig. 5. In the initial position, the stand bearing the sample is at the bottom,

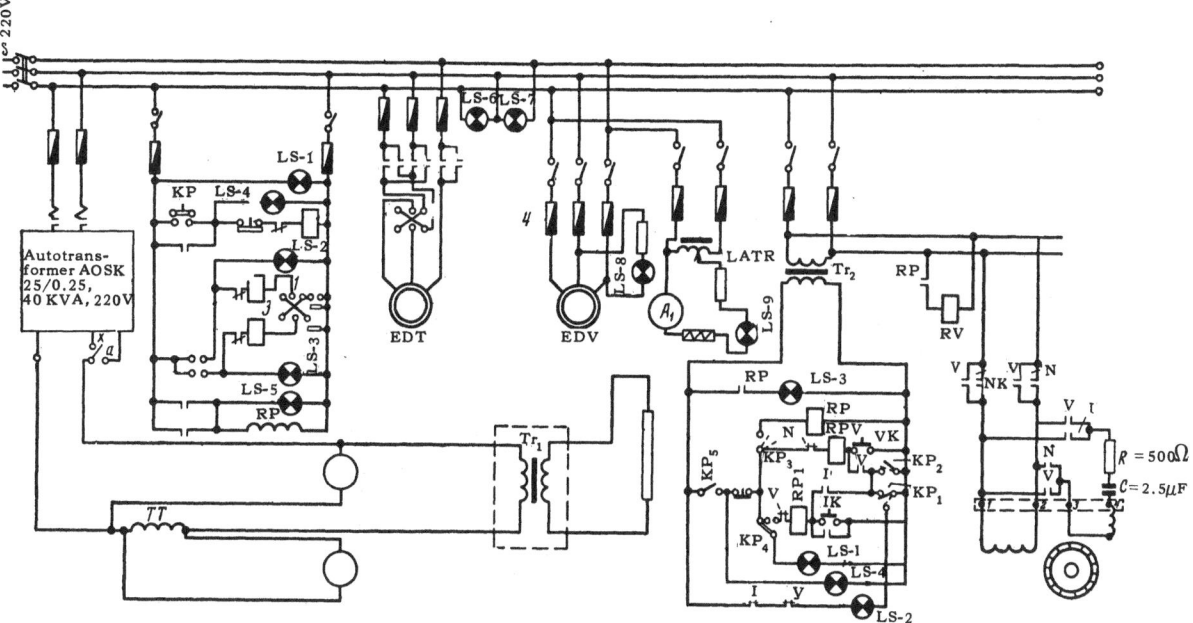

Fig. 5. Electrical circuit of semi-automatic control system for hot-hardness measuring apparatus.

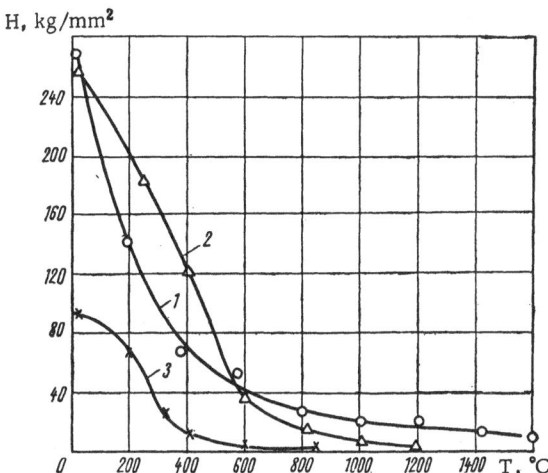

Fig. 6. Hardness as a function of temperature: 1) Annealed molybdenum; 2) nickel; 3) copper.

the terminal swtiches KP_4 and KP_2 are open, and KP_1 is closed; a white light LS-1 shows, and the motor is off. The motor is set to lift the stand by moving the tumbler from position NK to VK. Thereupon the winding of relay RPV is switched on, closing all the contacts V; the tumbler VK is shunted through the contact and the switch KP_1, so that the motor continues to operate even after the dropping of the tumbler. When the stand enters the heating zone, KP_1 opens and the motor is switched off; a red light LS-2 then shows. After the holding period in the heating zone, a second move of the tumbler to the VK position causes the motor to raise the stand to the testing zone. Switch KP_2 closes (KP_1 remaining open) and the tumbler is shunted, but the motor remains switched on even after the dropping of the tumbler.

When the sample has received the load, terminal switch KP_3 opens the motor circuit and simultaneously closes the time relay RV through the intermediate relay RP. The time spent by the sample under the load is determined by previously setting the relay RV; for all this period the green light LS-3 shows.

When the holding period ends, the relay RV switches the motor on in the reverse direction and the stand is let down out of the test zone. When the stand reaches the heating zone, switch KP_2 opens and the stand stops.

Subsequent operations depend on the nature of the tests:

a) For "impression" tests, the sample is rotated through $1/13$ of a turn, lifted to the testing zone, and the whole cycle repeats;

Fig 7. Apparatus for measuring hot hardness.

b) for "flattening" tests, the first sample is replaced by another, after letting the stand with the sample down to the initial position; for this the tumbler is set to position NK, which closes switch KP_1; when the stand reaches the bottom position, KP_4 opens and the stand stops.

In order to avoid accidents to the loading mechanism, the circuit includes a blocking terminal switch KP_5; closing this by the master switch of the loading system switches the circuit on. The position "stand ready to lift" is indicated by signal light LS-4. In this position the loading system remains during the whole test.

Vacuum System. In order to obtain a vacuum of 10^{-4} to 10^{-5} mm Hg in the apparatus, we use a backing pump followed by a diffusion pump. The chamber is pumped out to a rough vacuum of 10^{-3} mm Hg by a VN-2 pump through a diffusion pump of the Ts-VL-100 type. The vacuum is measured by means of a VIT-1P vacuum gage and measuring tubes of the LT-2 and LM-2 types. The vacuum system is monitored and regulated from the desk in front of the equipment.

Temperature Control. The temperature is controlled in two ways: with an OP-48 optical pyrometer graduated at 5000°C and with three thermocouples (Chromel-Alumel, tungsten-molybdenum-aluminum, and tungsten-rhenium). In order to obtain accurate data when using the optical pyrometer, corrections are made for absolute black-body conditions and for absorption of radiation by the glass of the observation window.

Testing the Apparatus and Some Typical Results

Tests of the experimental system showed that for temperatures up to 1600°C all units operated satisfactorily. In preliminary experiments we measured the hot hardness of molybdenum, nickel, and copper samples (Fig. 6).

The reliability of the apparatus (a general view of which appears in Fig. 7) is still being checked.

Literature Cited

1. Borzdyka, A. M. Methods of Carrying out Hot Mechanical Tests on Metals, Moscow, Metallurgizdat, 1962.
2. High-Temperature Strength in Machine Construction. Transactions of the Scientific and Technical Congress in Kiev, 1962, p. 230.
3. Fitzgerald, L. M. Brit. J. Appl. Phys., 11(12):(1960).

APPARATUS FOR DYNAMIC TESTS ON METALS
AND ALLOYS IN ORGANIC HEAT CARRIERS
AT HIGH TEMPERATURES AND PRESSURES

Yu. F. Bychkov, I. D. Laptev, and A. N. Rozanov

For testing the corrosion resistance of metals and alloys in organic heat carriers, the materials must be subjected to conditions approaching those found in practice. For this purpose various agitating systems have been devised so as to enable dynamic tests to be carried out under laboratory conditions.

Various schemes for dynamic tests in agitators have been published [1]. The disadvantage of these is that it is difficult accurately to determine the velocity at which the samples move relative to the corrosive media, since there is always a fall in velocity within the boundary zone around the samples. In addition to this, in some systems an additional centrifugal force acts on the sample [2]. Only a few samples are tested at one time in the agitator.

We have designed a system for testing the corrosion resistance of metals and alloys in organic heat carriers at temperatures up to 400 or 450°C, pressures up to 50 atm, and computed velocities between 1.4 and 5.7 m/sec. The test conditions are considerably closer to the practical conditions in which metals and alloys operate, i.e., the heat carrier moves and the samples are at rest. No cavitation or centrifugal forces occur. This apparatus constitutes a small-scale system for testing the corrosion of metals and alloys in organic heat carriers.

Our system is distinguished from others designed for the same purpose by its much smaller dimensions, by being much less complicated to set up, adjust, and use, and by the fact that the tests require only a small quantity of heat carrier (0.3 liter) as compared with tens of liters required by the earlier systems; this is important when experimental samples of organic heat carriers are being tested.

The apparatus (Fig. 1) proposed for long corrosion tests in an organic heat carrier consists of the following principal components: drive, finned neck, and autoclave. Inside the autoclave is a screw for driving the liquid and a distributor which "disentangles" the flow of liquid thrown out by the screw and thus creates a static thrust as required for the circulation of the heat carrier.

The distributor is covered with a cassette consisting of inner and outer holders (Fig. 2). The inner holder has 20 longitudinal channels of identical cross section, 5 by 8 mm, set at equal distances around the circumference. In the channels of the outer and inner holders are 20 rows of slits, two to a row. These are intended for fixing samples 40 × 10 × 1 mm in size made of

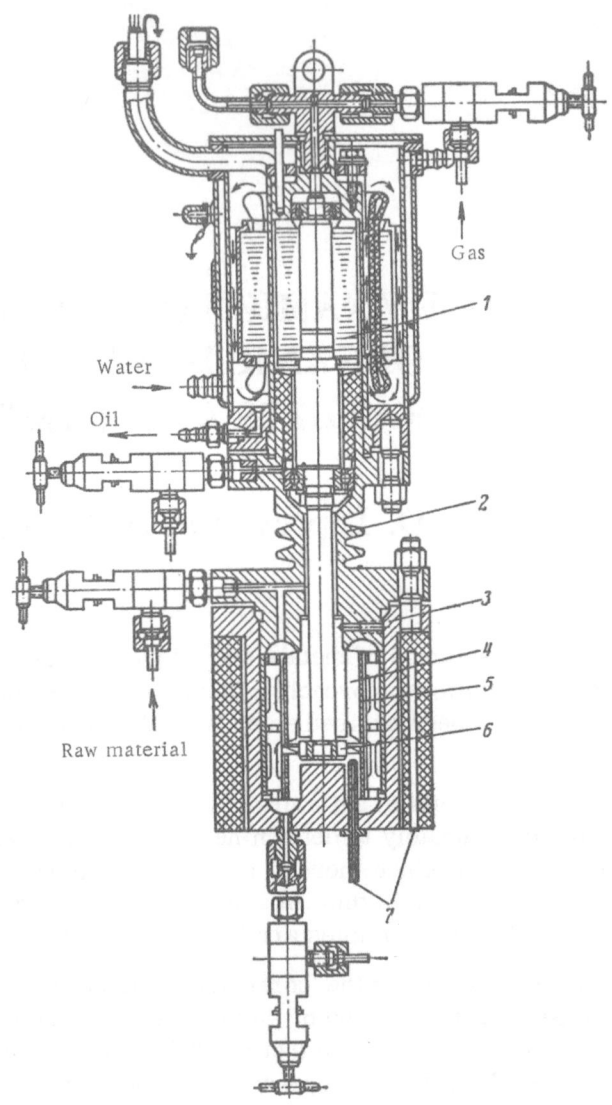

Fig. 1. Apparatus for dynamic tests on metals and alloys in organic heat carriers at high temperatures and
pressures: 1) Drive; 2) neck; 3) autoclave; 4) distributor; 5) cassette; 6) screw; 7) pocket for thermocouple.

sheet material along the axis of the channels. The heat carrier moves along the channels and
flows around the samples on two sides, exactly along their surfaces.

According to [3], the velocity of the heat carrier in the diffuser is

$$U_d = \eta D n P,$$

where η is the nominal efficiency of the screw, D is the screw diameter, n the number of
turns of the screw, and P the pitch ratio.

In determining the velocity of the heat carrier around the sample, however, we must allow for the ratio of the free cross-sectional area in the plane of the screw to the cross-sectional area of the annular space. Then, in general form, the velocity of the heat carrier around
the sample is

Fig. 2. Cassette for fixing
the samples.

Fig. 3. Distributor, screw,
and cylinder.

Fig. 4. Apparatus for dynamic tests.

$$U = \eta D n P \frac{\pi (D^2 - d^2)}{4k (F - 8t)} ,$$

where d is the diameter of the bushing, k the number of channels (k = 20), F is the channel cross section, and t the thickness of the sample.

The velocity of the liquid along the samples is varied by changing the screw to another pitch ratio.

With this fixing arrangement, the samples experience no extraneous loadings due to the rotation of the liquid or the sample; there is also no cavitation such as that which occurs when the samples are rotated. In order to prevent the 40 samples from falling out, a cylinder is placed around the cassette (Fig. 3). All components in contact with the organic substances are made of 1Kh18N9T stainless steel, which has a high corrosion resistance to these liquids.

The drive for the apparatus is of the kind used in the laboratory isothermal reactor R-A5-31 (developed in the Leningrad Branch of the Scientific-Research Institute of Chemical Engineering under the direction of N. E. Vishnevskii) intended for the hydrogenation of hydrocarbons and fats, as well as other processes taking place at high pressures and requiring intense circulation of the reacting substances.

Valves are provided for evacuating the apparatus before operation, filling it with inert gas, pouring in 0.3 liter of organic heat carrier, and running off after the tests.

The apparatus for dynamic tests on metals and alloys in organic heat carriers at high temperatures and pressures shown in Fig. 4 was designed, manufactured, and successfully tested under the following conditions: temperature 320°C, pressure 8 atm, organic heat carrier monoisopropyldiphenyl with 0.1% water, operating period 500 h.

Literature Cited

1. Balandin, Yu. F., and Markov, V. G. Construction Materials for Apparatus Containing Molten-Metal Heat Carriers, Moscow, Sudpromgiz, 1961.
2. Pearlman, H. Proceedings of a Symposium Held in Vienna, May 10-13, 1960, Vol. 1, pp. 203-204.
3. Vishnevskii, N. E., et al. High-Pressure Apparatus with a Hermetic Drive, Moscow, Mashgiz, 1960, p. 156.

RADIATION HEATER FOR FLOATING–ZONE REFINING

I. V. Milov, D. M. Skorov, and V. V. Okinshevich

In floating-zone refining decisive factors are the stability of the molten zone (fully discussed in [1, 2]), the temperature distribution in the sample, and the manner in which the incoming thermal flux required to create a constant-length zone varies with the position of the zone in the sample. The last two points were fully considered in [3–7]. A simple and convenient method of creating a molten zone using a floating-zone technique is radiation heating of the zone from an annular radiator [8]. Such a radiator may be heated to the required temperature T_K (°K) by high-frequency currents [9] or the direct passage of a current through the heater. This kind of heating is suitable for materials having a poor coupling with the electromagnetic field of the inductor, and also for improving the shape of the molten-zone boundary in induction-type floating-zone refining.

Radiation heaters are especially convenient in operation, enabling the thermal conditions of the zone-refining process to be readily controlled. For high-frequency heating, the instability of the generator output power has little effect on the size and shape of the zone owing to the stabilizing action of the ring. Radiation heaters nevertheless have the following disadvantages:

1. At the high temperatures required to produce high emission of the radiator, the material composing the latter may evaporate and contaminate the substance being studied.

For this reason the most suitable materials for radiators are tantalum, tungsten, and carbon, which have very high melting points and low vapor pressure at temperatures up to 2400 or 2500°C at which radiators usually operate.

2. On using induction heating, there is a certain screening of the sample from the field of the inductor. Hence mixing in the zone is reduced and the efficiency of purification falls.

Despite these disadvantages, the use of annular radiators is quite promising.

In this paper we shall describe a method of designing such a radiator, together with the results of experimental tests. By using the formulas of [3–7] to calculate the power required to create a stable zone of given length, we can determine the temperature and dimensions of the radiator (ring). These formulas relate the thermal-flux and power distribution along the sample to the geometric dimensions of sample and radiator and also to the radiator temperature.

We carried out calculations for the vacuum zone refining of a cylindrical sample $2r$ in diameter by an annular radiator $2a$ wide arranged concentrically with the sample (Fig. 1). The inner diameter of the radiator was $2R_1$, the external diameter $2R_2$, the temperature T_K, the radiating power of the radiator material A_K, the absorption coefficient of the sample material A_1. We consider that the material of the ring radiates on a cosine law, i.e., isotropically. The emission from the radiator is given by the equation

$$N_0 = A_\kappa \sigma \cdot T_\kappa^4,$$

where σ is the Stefan-Boltzmann constant.

Fig. 1. Arrangement of annular radiator and sample: r) Radius of sample; R_1 and R_2) inner and outer radii of radiator; 2a) width of radiator; 1) sample; 2) radiator.

Let us consider the distribution of flux density from the inner surface of the annular radiator over the sample. We shall suppose that the area of surface $dS_1 = R_1 d\varphi \cdot dt$ radiates as a point source lying at its center (Fig. 2). This emits a flux of $N_0 dS_1$ into a solid angle of 2π; the proportion of this flux falling on the sample is $N_0 dS_1(\varphi_0/\pi)$, where $\varphi_0 = 2\sin^{-1}(r/R_1)$. Owing to the circular symmetry of the problem, we may consider that the whole of this flux falls on a section of sample $rd\varphi$ and passes inside an angle $d\varphi_1$. In this approximation, the problem may be regarded as one-dimensional. Then the flux proceeding in direction i equals

$$N_0 dS_1 \frac{\varphi_0}{\pi} \cos i \cdot \frac{1}{2}. \tag{1}$$

The factor $1/2$ allows for the fact that the flux in direction i may travel in two opposite directions. The flux falling on area dx equals

$$N_0 R_1 \frac{\varphi_0}{2\pi r} \frac{\cos^2 i}{R} d\tau dS_2, \tag{2}$$

where $dS_2 = r \cdot d\varphi \cdot dx$ is the area of the sample and $R = [(x-t)^2 + (R_1-t)^2]^{1/2}$.

The flux density along the sample can be obtained by integrating expression (2) (after dividing by dS_2):

$$I_{in} = N_0 R_1 \frac{\varphi_0}{2\pi r} \int_{-a}^{a} \frac{\cos^2 i}{R} d\tau = N_0 R_1 \frac{\varphi_0}{2\pi r} \left[\frac{a-x}{[(a-x)^2 + (R_1-r)^2]^{1/2}} + \frac{a+x}{[(a+x) + (R_1-r)^2]^{1/2}} \right]. \tag{3}$$

As R_1 tends to r, $\varphi_0 = 2\sin^{-1}(r/R_1)$ tends to π, and

$$\frac{a-x}{[(a-x)^2 + (R_1-r)^2]^{1/2}} \rightarrow \begin{cases} 1 & \text{for } \infty < x \leqslant a \\ -1 & \text{for } x > a \end{cases}$$

$$\frac{a+x}{[(a+x)^2 + (R_1-r)^2]^{1/2}} \rightarrow \begin{cases} 1 & \text{for } -a \leqslant x < \infty \\ -1 & \text{for } x < -\infty \end{cases}.$$

The sum of these two terms tends to 2 for $-a \leq x \leq a$ and to 0 for $x > a$ and $x < -a$;

Hence as $R_1 \rightarrow r$, $I_{in}(x) \rightarrow N_0$ for $-a \leq x \leq a$ and $I_{in}(x) \rightarrow 0$ for $x > 0$, $x < -a$.

This result proves the validity of the derivation of equation (3).

We may solve the problem for the side surface of the ring analogously. In this case, an element of ring surface $yd\varphi dy$ transmits a flux $N_0 dS_1(\varphi_S/2\pi)$, where $\varphi_S = 2\sin^{-1}(r/y)$ (Fig. 3). The flux proceeding in direction i equals $N_0 dS_1(\varphi_0/2\pi)\sin i$. The flux falling on area dx equals

$$\frac{N_0}{2\pi r} \cdot \frac{y\varphi_S \cdot \sin i \cdot \cos i \, dy}{R} dS_2. \tag{4}$$

The flux density over the length of the sample equals

$$I_S(x) = \frac{N_0}{2\pi r} \int_{R_1}^{R_2} \frac{y \cdot \varphi_S \cdot \sin i \cdot \cos i}{R} dy = \frac{N_0}{\pi r} \int_{R_1}^{R_2} \frac{y \sin^{-1} \frac{r}{y} (y-r)(x-a)}{[(x-a)^2 + (y-r)^2]^{3/2}} dy, \tag{5}$$

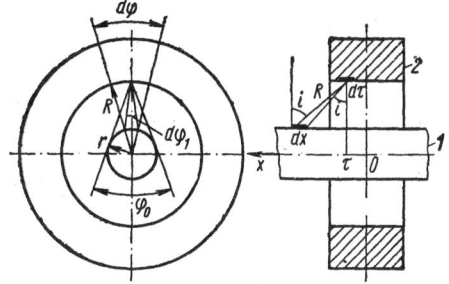

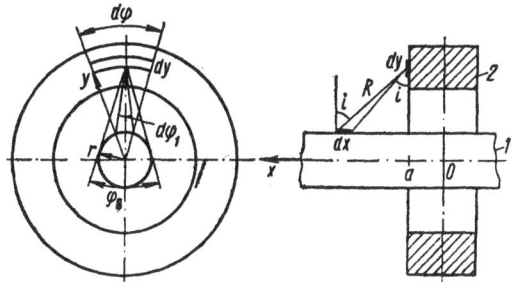

Fig. 2. To illustrate the calculation of the thermal-flux density distribution along the sample arising from the inner surface of the radiator: 1) Sample; 2) radiator.

Fig. 3. To illustrate the calculation of thermal-flux distribution along the sample from the side surface of the radiator: 1) Sample; 2) radiator.

where $\sin i = (x-a)/R$, $\cos i = (y-r)/R$, $R^2(x-a)^2 + (y-r)^2$; the integral on the left of Eq. (5) cannot be integrated in the form of rational functions. The integral may however be calculated to any degree of accuracy by expanding arc sin (r/y) in series, the number of terms taken depending on the ratio r/R_1 and the accuracy required.

To a fair degree of accuracy we may replace arc sin (r/y) by φ_{av}, equal to $\dfrac{\sin^{-1}\dfrac{r}{R_1} + \sin^{-1}\dfrac{r}{R_2}}{2}$. This substitution only affects the result at small $(x-a)$, but here $I_S(x)$ is also not large. Hence the error introduced by the substitution is small.

After making all the substitutions and integrating Eq. (5), we obtain

$$I_s(x) = \frac{R_1}{[(x-a)^2 + (R_1-r)^2]^{1/2}} - \frac{R_2}{[(x-a)^2 + (R_2-r)^2]^{1/2}} + \ln \frac{R_2 - r + \sqrt{(x-a)^2 + (R_2-r)^2}}{R_1 - r + \sqrt{(x-a)^2 + (R_1-r)^2}}. \qquad (6)$$

The total flux is

$$I(x) = I_{in}(x) + 2I_s(x). \qquad (7)$$

Now we shall consider the total power imparted to a sample of length $2l$; this equals $\int_{-l}^{l} I(x)\,dx$.

Only flux traveling in the dihedral angle φ_0 falls on the sample from the inner surface of the ring. From this we must subtract flux traveling in the solid angles α_{in} (to right and left) (Fig. 4).

Taking $2a \ll 2l$, we find that a flux

$$N_0 = 2\pi R_1 \cdot 2a \cdot \frac{\varphi_0}{\pi}\left[1 - \frac{2(R_1-r)}{\pi \cdot l}\right], \qquad (8)$$

falls on the sample, since $R_1 - r \ll l$.

The flux from the side surface is calculated as

$$dI_s = N_0 \cdot 2\pi y \cdot \frac{\varphi_s}{2\pi} \cdot d\varphi \cdot \frac{\varphi_y}{\pi/2}, \qquad (9)$$

where $\varphi_y = \cot^{-1}\dfrac{y-r}{l-a}$ (see Fig. 3).

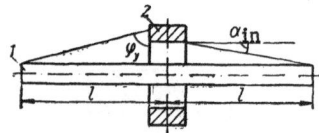

Fig. 4. To illustrate the power imparted to the sample: 1) Sample; 2) radiator.

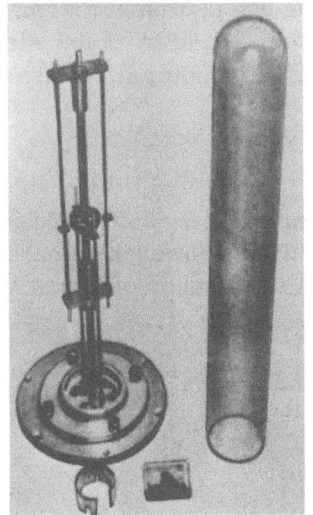

Fig. 5. Top part of vacuum chamber, with sample, screen, and quartz envelope.

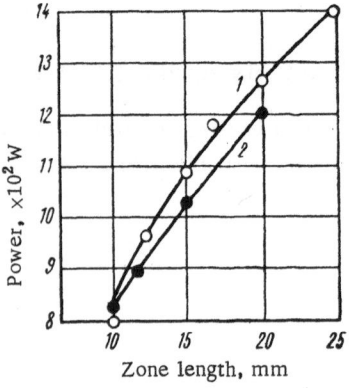

Fig. 6. Length of a zone situated symmetrically with respect to the ends of the sample as a function of the power conveyed to the zone: 1) Experimental curve; 2) theoretical curve.

Integrating Eq. (9), we obtain

$$I_s = N_0 \frac{4\varphi_{av}}{\pi} \int_{R_1}^{R_2} y \, \cot^{-1} \frac{y - r}{l - a} \, dy$$

$$= N_0 \frac{2\varphi_{av}}{\pi} \left\{ \frac{\pi}{2} (R_2 - R_1)(R_2 + R_1) - \right.$$

$$- \frac{(R_2 - r)^2 (R_2 + r) - (R_1 - r)^2 (R_1 + r)}{l - a}$$

$$\left. + r(l - a) \frac{(R_2 - r)^2 - (R_1 - r)^2}{(l - a)^2 + (R_1 + r)^2} \right\}. \tag{9'}$$

If we neglect the quantity $(R_1 - r)^2$ as compared with $(l - a)^2$, then Eq. (9') simplifies:

$$I_s = N_0 \frac{2\varphi_{av}}{\pi} \left\{ \frac{\pi}{2} (R_2^2 - R_1^2) - \frac{(R_2 - r)^2 R_2 - (R_1 - r)^2 R_1}{l - a} \right\}. \tag{10}$$

The second term in the brackets in Eq. (10) takes account of the finite length of the sample; for $l \to \infty$ it drops out.

The total power falling on the sample is

$$I = I_{in} + 2I_s. \tag{11}$$

The total power N given to the sample may be determined if we multiply I by the absorption coefficient of the metal forming the sample, equal to the emissivity of the metal A_c

$$N = I \cdot A_c. \tag{12}$$

It was shown in [3] that a zone 2b long has plane boundaries if $I(b) = A_c \sigma T_m^4$ at $x = b$ (T_m = melting point of the sample material). By using this condition and Eq. (12), we may calculate the optimum parameters of the radiator. The outer diameter of the ring has little influence on the result.

We checked the validity of the formulas here derived with an apparatus designed for the floating-zone refining of beryllium. In the first form of the apparatus, the beryllium sample was heated by the radiation of a tantalum ring, itself heated by high-frequency currents (Fig. 5). This method of heating was used for the floating-zone refining of beryllium in [10]. In the second form of the apparatus, the heat source was an open annular heater concentric with the sample and heated by a current passing directly through it.

Figure 6 shows the length of a zone situated symmetrically with respect to the ends of the sample as a function of the power directed into the zone (curve 1); the theoretical curve was calculated from our own formula (12) and formula (3) of [3].

The experimental and theoretical results are in satisfactory agreement.

Conclusions

1. A method of calculating the parameters of a radiation heater for zone recrystallization is set out.

2. An experimental proof of the method is given for the zone recrystallization of beryllium; the theoretical and experimental curves are in satisfactory agreement.

Literature Cited

1. Heywang, W. Z. Naturforsch., 11a:238 (1956).
2. Heywang, W., and Liegler, G. Z. Naturforsch., 9a:561 (1954).
3. Milov, I. V., et al. Izv. Akad. Nauk SSSR, Otd. Tekhn. Nauk, Met. i Toplivo, No. 2:56 (1962).
4. Volchok, B. A., and Frenkel', V. Ya. Inzh.-Fiz. Zh., 4:43-48 (1961).
5. Donald, D. K. Rev. Sci. Inst., 32:811 (1961).
6. Ratnikov, D. G., et al. Prom. Primenenie Tokov Vysokoi Chastoty Elektro., 1961, p. 124.
7. Zabelina, L. G., et al. Inzh.-Fiz. Zh., 5:81 (1962).
8. Pfann, W. G. Zone Melting [Russian translation], Moscow, IL, 1961. [English edition: Wiley, New York, 1958.]
9. Emeis, R. Z. Naturforsch., 9a:67 (1954).
10. Edwards, K. L., and Martin, A. Transactions of the International Conference on the Metallurgy of Beryllium, London, October 16-18, 1961 [Russian translation].

SUMMING DEVICE FOR AN X-RAY DIFFRACTION CAMERA

G. A. Mochalov

In studying annealing texture by the x-ray method, it is necessary to obtain x-ray diffraction pictures from coarse-grained samples. In this case only a smallish number of grains lie in the reflecting position. In order to be able to judge the texture, the number of such grains must be increased. For this purpose various summing devices are normally used [1-6].

In the majority of cases, new cameras are designed [2-4] or fairly complicated attachments are made [1]. We have tested the operation of a simple summing device for the industrial RKSO x-ray camera; this avoids altering the camera itself.

The general view of the summing device, with the camera arranged on the operating table of the URS-70 system, is shown in Fig. 1; the corresponding design diagram is shown in Fig. 2. Rotation is imparted to a steel cam, with a shaft coming out of the body through a side cut, from an SD-2 motor via a flexible metal cable. On rotation of the cam in a brass holder, smooth to-and-fro motion of a brass cylinder in a steel housing automatically takes place through a distance of some 4 mm in a vertical direction. A spring ensures good contact of the cylinder with the cam and prevents possible slowing of the cylinder in its downward motion. A guiding screw occupying a groove in the cylinder prevents rotation of the latter around its own axis.

At the top of the movable cylinder, guiding and adjusting plates providing freedom of movement for the stand are fixed in a definite position relative to the film cassette by means of a screw (Fig. 2). The housing is screwed to the support column, and this, carrying the summing device in assembled form, is placed in the RKSO camera instead of the usual goniometer-head holder, where it is rigidly fixed.

A plane sample of the material being studied is placed on the stand with the help of modeling clay in such a way that it can be moved in its own plane (through a maximum distance of 12 mm) on moving the stand by turning the shaped screw.

Thus the total area of irradiation of the sample is approximately 50 mm^2 (12 × 4).

The primary x-ray beam falls perpendicularly to the sample surface; the original rolling direction is vertical. Subsequent rotations of the whole summing system through discrete angles around the rolling direction (which coincides with the camera axis) are made by means of the graduated circle of the RKSO camera. The side cut in the housing, and the holder freely rotating in the support, enable the sample to be rotated up to 150° around the camera axis without moving the SD-2 motor relative to the camera. The motor is placed freely on the table of the x-ray apparatus.

While photographs are being taken with the plane of the sample at any arbitrary angle to the incident beam, there is a continuous displacement of the sample in its own plane, and the

Fig. 1. General view of the summing device
for the RKSO x-ray diffraction camera.

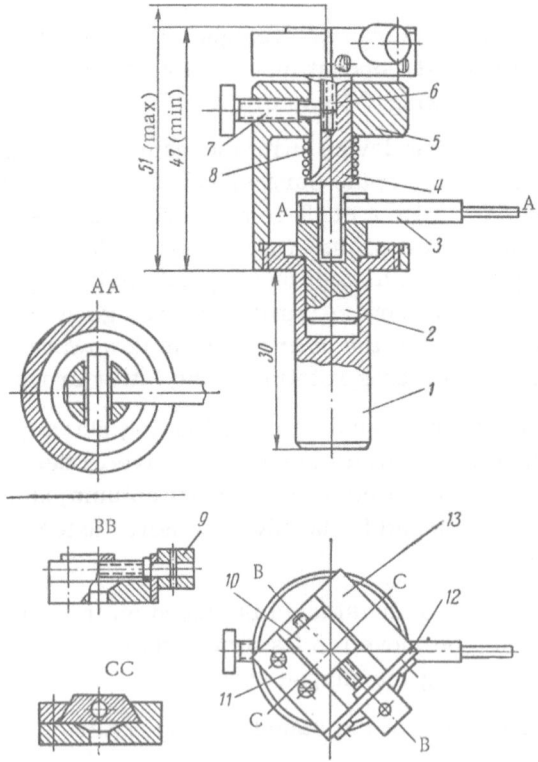

Fig. 2. Principle of summing device for the RKSO
x-ray diffraction camera: 1) Support; 2) holder; 3)
cam with shaft; 4) movable cylinder; 5) housing; 6)
fixing screw; 7) guiding screw; 8) spring; 9) shaped
screw; 10) stand; 11) adjusting plate; 12) centering
plate; 13) guiding plate.

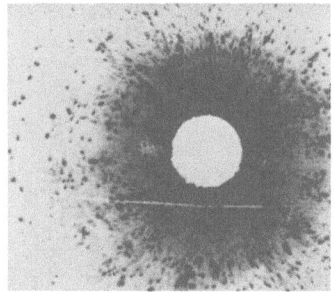

Fig. 3. X-ray photograph of
an industrial molybdenum foil
after annealing at 1400°C for
1 h in a vacuum of 10^{-5} mm Hg
(no summing device used). Exposure time 20 min.

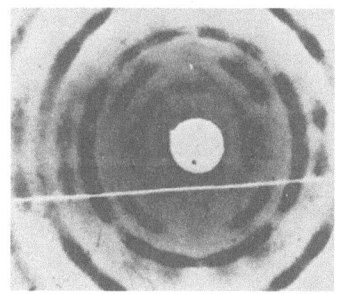

Fig. 4. X-ray diffraction
picture of the industrial
molybdenum foil shown in
Fig. 3, taken with the aid of
the summing device. Exposure
time 3 h.

sample may periodically be moved horizontally in the same plane by means of the manually controlled shaped screw.

Figures 3 and 4 show x-ray diffraction pictures obtained from an annealed molybdenum
foil (99.9% purity) 50 μ thick. The photographs were taken in filtered molybdenum radiation
with a voltage of 45 KV and current 12 ma. Figure 3 (no summation) shows a random distribution of x-ray reflections from recrystallized grains (grain size around 0.05 to 0.08 mm). The
photograph taken with the summing device (Fig. 4) gives a picture with a regular distribution
of maxima. In this case we have a texture photograph of a completely recrystallized molybdenum foil. Clearly this summing system may also be used for other organic and inorganic
recrystallization textures.

Literature Cited

1. Borodkina, M. M. Zavod. Lab., 26:491 (1960).
2. Izbranov, P. D., et al. Vestn. Vyssh. Shkoly, No. 6:84 (1960).
3. Fink, W., and Smith, D. W. Symposium on Radiography and X-Ray Diffraction, ASTM,
   1937, p. 200.
4. Wilson, F. H., and Brick, R. M. Am. Inst. Min. and Metallurg. Engng., 161:173 (1945).
5. De Barr, A. E., and Roberts, B. J. Iron and Steel Inst., 164:287 (1950).
6. Guipier, A. Rev. Metallurgie, 45:277 (1948).

MELTING DIAGRAM OF THE SYSTEM FORMED BY A MIXTURE OF RARE-EARTH CHLORIDES AND SODIUM CHLORIDE

K. V. Orlov, V. G. Kozlov, and N. G. Pospelova

This paper contains information on interaction between the mixed chlorides of a number of rare-earth elements of the cerium group and sodium chloride in the form of melts. The question is of special interest in connection with the retreatment of rare-earth raw materials by the chlorination method and in producing rare-earth metals by the electrolysis of molten media.

A number of systems formed by alkali and alkaline-earth metals with the chlorides of individual rare-earth elements (r.e.) have been studied, both in the Soviet Union and elsewhere [1-9]. There are nevertheless no published data regarding the system formed between sodium chloride and a sum of r.e. chlorides.

The r.e. chloride (cerium group) required for this purpose was obtained as follows: A portion of the original r.e. oxide was dissolved in excess of hydrochloric acid; the resultant solution was filtered and evaporated in the presence of ammonium chloride until the crystal hydrates separated out.

Then the water, excess acid, and ammonium chloride were distilled off at 250 to 300°C in a vacuum system at a residual pressure of 30 to 40 mm Hg.

The resultant sum of r.e. chlorides had the following composition:

$$\begin{aligned} CeCl_3 &- 53\% \\ LaCl_3 &- 30\% \\ NdCl_3 &- 8\% \\ PrCl_3 &- 3\% \\ RE'Cl_3 &- 6\% \end{aligned}$$

where $RE'Cl = \Sigma SmCl_3$, $EuCl_3$, and other r.e.

We found the melting point of the mixture to be 780°C.

We used this form of RE'Cl and sodium chloride of analytical purity to obtain the melting diagram.

The sodium chloride was previously dehydrated at 600°C for 2 to 3 h.

The interaction of the components in the system was studied by the fusion method, the cooling curves being recorded with an FPK-55 Kurnakov pyrometer.

Results of a Thermal Analysis of the $\Sigma RECl_3$ —NaCl System

Content				Temperature, °C	
wt. %		mol. %			
$\Sigma RECl_3$	NaCl	$\Sigma RECl_3$	NaCl	first effect	second effect
100,0	—	100.0	—	780.0	—
90,0	10.0	68.0	32.0	668.0	484.0
80,0	20.0	48.5	51.5	558.0	484.0
70,0	30.0	35.5	64.5	488.0	Eutectic
60,0	40.0	26.0	74.0	558.0	488.0
50,0	50.0	19.5	80.5	636.0	488.0
40,0	60.0	13.7	86.3	712.0	488.0
30,0	70.0	9.1	90.9	748.0	488.0
20,0	80.0	5.6	94.4	776.0	488.0
10,0	90.0	2.6	97.4	784.0	488.0
—	100.0	—	100.0	800.2	—

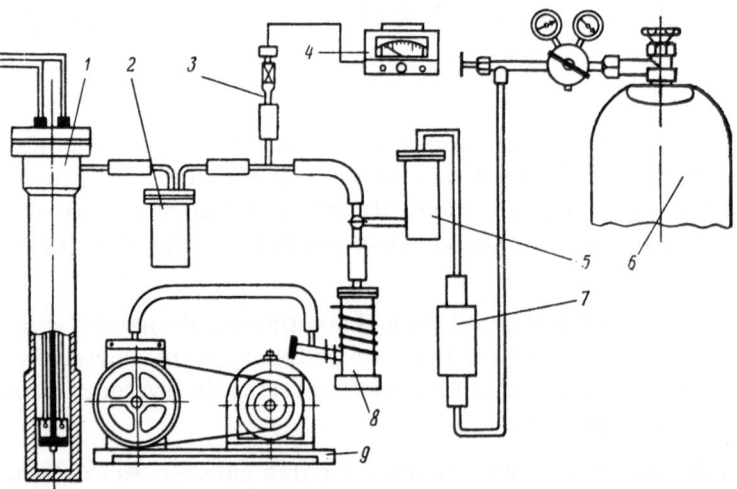

Fig. 1. Arrangement of apparatus for thermal analysis: 1) Retort; 2, 5) traps; 3, 4) vacuum-measuring system; 6) argon cylinder; 7) furnace containing copper filings; 8) diffusion pump; 9) backing pump.

In view of the fact that the r.e. chlorides decompose readily at high temperatures in the presence of oxygen and are subject to hydrolysis on interaction with atmospheric moisture, the salt mixtures were melted in an argon atmosphere, using the apparatus shown in Fig. 1.

The rate of cooling the furnace containing the sample was 8 to 10 deg/min. The temperature was measured with a Pt-PtRh thermocouple.

The results of the thermal analysis are presented in the table and Fig. 2.

The resultant melting diagram of the $\Sigma RECl_3$ —NaCl system is of the eutectic type. The eutectic contains 35.5 mol.% $\Sigma RECl_3$ and melts at 488°C.

Conclusions

1. We have studied the $\Sigma RECl_3$ —NaCl system by differential thermal analysis (RE indicates the sum of r.e. of the cerium group in natural ratio as obtained from loparite).

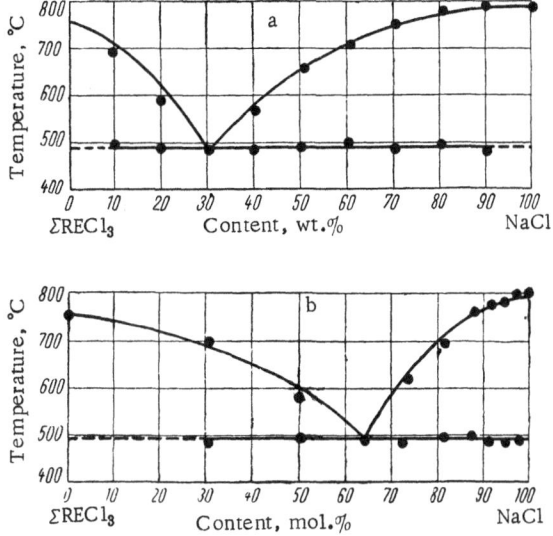

Fig. 2. Phase diagram of the $\Sigma RECl_3$—NaCl system.

2. The system melts on a simple eutectic principle.

The composition of the eutectic is 35.5 mol.% $\Sigma RECl_3$ (70 wt.%) and the melting point 488°C.

Literature Cited

1. Handbook on the Fusion of Salt Systems, Vol. 1, Moscow, Izd. Akad. Nauk SSSR, 1961.
2. Nishihara, K., et al. Electrochem. Soc., Japan, 19:105 (1951).
3. Nishihara, K. J. Electrochem. Soc. Japan, 18:179 (1960).
4. Nishihara, K., et al. Bull. Inst. Chem. Res. Kyoto Univ., 29:81 (1952).
5. Pushin, N. A., and Baskov, A. V. Zh. Russk. Fiz.-Khim. Obshch., 45:82 (1913).
6. Dergunov, E. P. Dokl. Akad. Nauk SSSR for 1948 to 1952.
7. Vogt, F. Neues Jahrb. Mineral. Geol. Palaontle, 9:2 (1914).
8. Morozov, I. S., et al. Zh. Neorgan. Khim., 7:1639 (1957).
9. Vereshitina, R. P., and Luzhnaya, N. P. Izv. Sektora Fiz.-Khim. Analiza, Inst. Obsch. Neorgan. Khim., Akad. Nauk SSSR, 25:188 (1954).